A Pictorial Guide to British Ephemeroptera

By Craig Macadam
and Cyril Bennett

FSC
BRINGING
ENVIRONMENTAL
UNDERSTANDING TO ALL

First edition © FSC 2010. OP139. ISBN: 978 1 85153 277 3

Acknowledgements

This guide has been produced by the Riverfly Partnership in Association with the Field Studies Council. The text and keys were developed by Craig Macadam, photographs by Cyril Bennett and line art by Rory McCann.

Special thanks are due to Rebecca Farley (Field Studies Council) for all her editorial advice; Bridget Peacock (Riverfly Partnership) for her encouragement and enthusiasm for this publication; and finally for all the people who tested this guide and provided valuable feedback.

Note

Due to the rarity of some species it has not been possible to get images of all the species covered in this guide. It is hoped that the missing gaps can be filled in the future so some blank pages have been left to accommodate this.

Contents

Acknowledgements	ii
Introduction	1
Identifying Ephemeroptera	2
Collection and Preservation	5
The Ephemeroptera Recording Scheme	5
Checklist of the British Ephemeroptera	6
Using this Guide	8
Identification charts and species accounts	
Ephemeroptera Families	
Identification Chart 1A – Nymphs	9
Identification Chart 1B – Adults	11
Baetidae	
Identification Chart 2A – Nymphs	13
Identification Chart 2B – Adults	14
Species Accounts	15
Caenidae	
Identification Chart 3A – Nymphs	42
Identification Chart 3B – Adults	43
Species Accounts	44
Ephemeridae	
Identification Chart 4A – Nymphs	60
Identification Chart 4B – Adults	61
Species Accounts	62
Ephemerellidae	
Identification Chart 5A – Nymphs	68
Identification Chart 5B – Adults	69
Species Accounts	71

Heptageniidae
    Identification Chart 6A – Nymphs … 74
    Identification Chart 6B – Adults … 75
    Species Accounts … 76

Leptophlebiidae
    Identification Chart 7A – Nymphs … 100
    Identification Chart 7B – Adults … 101
    Species Accounts … 102

Potamanthidae
    Species Account … 115

Siphlonuridae
    Identification Chart 8A – Nymphs … 116
    Identification Chart 8B – Adults … 117
    Species Accounts … 118

Adult fly identification and flight period tables … 124

References … 126

INTRODUCTION

Mayflies can be seen emerging from the water, resting on nearby vegetation and most commonly 'dancing' above head height, along the shores of many still waters and rivers. Also known as Ephemeroptera or up-wing flies, they can be identified by two large upright wings, two or three long tails and (in most species) two small hindwings. The word mayfly is misleading as this group of flies can appear throughout the year. The name comes from the habit of one species, *Ephemera danica*, which emerge as adults when the mayflower or hawthorn is in bloom. Around 3,000 species have been described from around the world and 51 of these species have been recorded from the British Isles. The mayfly has a fascinating life cycle, and is the only insect that has two adult stages: the sub-imago (or dun) and imago (or spinner).

The life cycle starts with the males forming a swarm and the females flying into the swarm to mate. The male grasps a passing female with its elongated front legs and the pair mate in flight, often falling to the ground in the process. After mating the female descends to the water surface to lay her eggs. Spent and motionless, the female lies with wings flat on the surface, often becoming prey to fish. The females of some species of Baetidae crawl under the water to lay their eggs directly on submerged stones. The male either rejoins the swarm or flies to nearby vegetation and dies. The eggs fall to the bottom of the water where they stick to plants and stones.

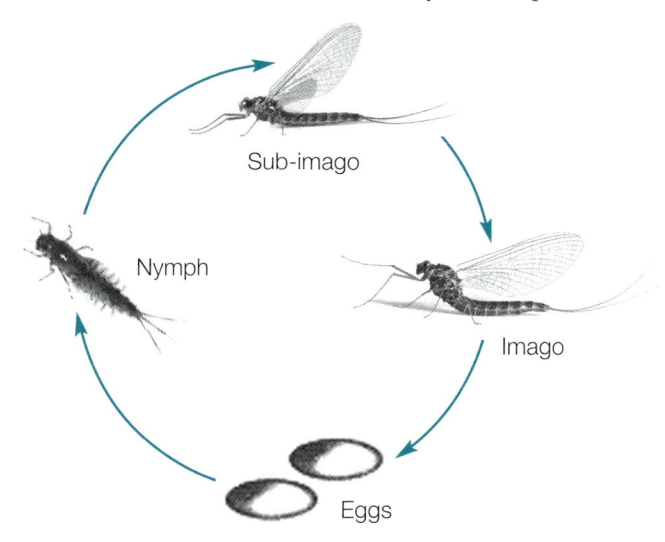

Over a period of a few days to a number of weeks, depending on water conditions and species, the eggs develop and hatch into tiny nymphs. The nymphs spend various lengths of time (up to two years) foraging on the bottom and pass through a number of moults (instars) before emerging as adults. When it is time to emerge, the nymph makes its way to the surface, pulls

Eggs of *Baetis*    Nymph of *Baetis*    Sub-imago of *Baetis*

itself free from the nymphal shuck (skin) and emerges as the dull-coloured sub-imago (or dun) and it is whilst resting on the surface of the water, to dry the newly exposed wings, that it is most vulnerable to attack from fish and birds. The sub-imago then seeks shelter in bank-side vegetation and trees. After a period of a couple of hours or more, the sub-imago once again sheds its skin to transform into the brightly coloured imago (or spinner). The function of the adult is to reproduce; the absence of mouthparts makes feeding impossible and so their remaining life is therefore dependent upon their stored energy reserves.

IDENTIFYING EPHEMEROPTERA

The identification of British (and Irish) Ephemeroptera is covered by two scientific publications by the Freshwater Biological Association (FBA) (Elliott & Humpesch, 1983; 2010, see page 126). These keys are primarily intended for use with preserved specimens and need a good binocular microscope, and in some cases a high power compound microscope, to identify your specimen. A third publication (Harker, 1989), in the Naturalists' Handbook series, is easier to use but still relies in places on the use of a high power microscope. This pictorial guide provides the information required to make a preliminary identification of most British Ephemeroptera. It also includes notes on their life cycle, ecology and distribution.

This guide is not intended to replace the FBA keys or Naturalists' Handbook 13, but instead to assist enthusiastic amateurs to attempt to identify up-wing flies. It is designed to be used primarily on live specimens, however to identify some species such as *Baetis rhodani* nymphs it is necessary to remove a gill for examination, which usually proves fatal. Most of the features used can be seen with the naked eye or with a x10 hand-lens. The use of magnification greater than this often hinders the identification process. There are cases within this guide where identification relies upon the detailed examination of the specimen with a microscope and users with suitable equipment are encouraged to consult the relevant FBA key for confirmation.

It is important when handling Ephemeroptera nymphs that you lift them by the legs. This prevents damage to the tails or gills that are important diagnostic features. This method also gives six opportunities to successfully lift the specimen!

It is often difficult to see the diagnostic features of Ephemeroptera. Anaesthesia can be used to slow down the movement of the gills of nymphs to allow an identification to be made. It is difficult to use in the field, but this method comes into its own in assisting examination under a microscope when you intend to keep the nymphs alive. The easiest, safest and cheapest method to anaesthetise nymphs is to place them into neat fresh soda water, or empty off the water in a collecting tube and replace it with soda water. Nymphs will survive anaesthesia for at least 15 minutes, but require a period of recovery in clean water afterwards.

Adults can be slowed down by subjecting them to low temperatures for a short period. At home this can be achieved by placing the specimens in the freezer compartment of a refrigerator for a couple of minutes. This method can also be used in the field by carrying a thermos flask filled with ice. The specimen is first placed in a watertight container and then placed in the flask for a short period.

Is it a mayfly?
Mayfly nymphs have three thread-like tails and up to seven pairs of gills along the body. These gills can be plate-like; strap-like; filamentous; or feather-like. Damselfly nymphs also have three tails however these are plate-like, rather than thread-like, and they never have external gills. Alderfly larvae have gills along the body but never have three tails, instead they have a single terminal filament.

Stonefly nymphs have only two tails and never have gills along the body, although in larger species there are filamentous gills under the legs and also under the tails. Whirligig beetle larvae have gills along the body but never have thread-like tails.

Adult mayflies are delicate insects with either two or three long, thread-like tails. They have relatively large forewings, often with many veins, which are held closed above the body like a butterfly when at rest. Mayflies may also have a pair of hindwings which are either relatively large and obvious, tiny and oval, thin and spur-like or absent.

Damselflies, lacewings and alderflies also have heavily veined wings which they hold above their body however they never have tails.

Is it a male or female?
When identifying larval mayflies there is no need to know the sex of the specimen, however with adults it is often useful to know which sex you are identifying. The principal difference between male and female adult mayflies is the presence of a pair of claspers below the tails of the male. In addition the male imago typically has larger eyes and longer front legs than the female.

Imago or sub-imago?
Mayfly sub-imagines have opaque wings fringed with small hairs and relatively short tails and legs. In the imago the tails and forelegs are much longer and the wings transparent and hairless. In the sub-imago the body is fully coloured, and this is also the case with the female imago. The body of the male imago is however almost transparent, apart from the last few segments which are fully coloured.

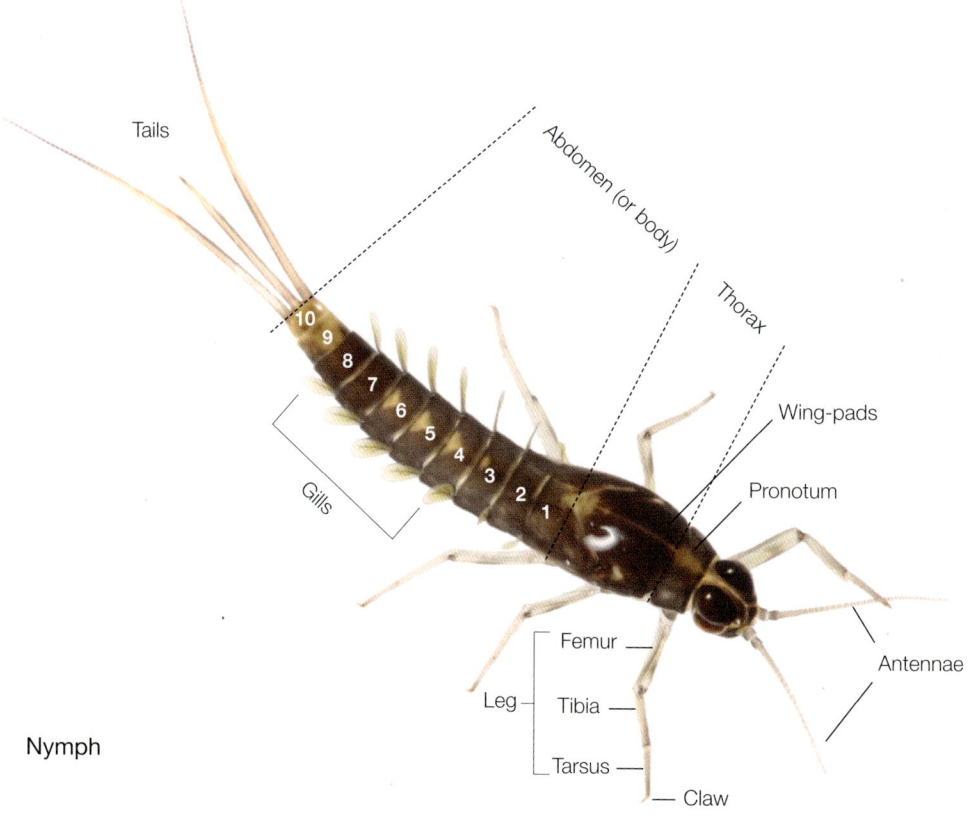

Main identification features used in this guide for adult and nymph.

COLLECTION AND PRESERVATION

Ephemeroptera nymphs are easy to collect. Kick sampling, the disturbance of the stream bed upstream of a collecting net, is the most efficient method in running water, while in standing water a net can be swept through submerged vegetation or the substrate can be disturbed and the net swept through the disturbed water. Adult Ephemeroptera can be collected by examining trees and other vegetation on the bank. Alternatively, adults can be caught as they swarm near the water.

This pictorial guide is designed to stimulate interest in identifying the British Ephemeroptera. Until you are familiar with this group, it is important to keep a **voucher specimen** to back up your identifications. Some species can be identified with confidence using this key, while with others a voucher specimen must be retained.

To prepare a voucher specimen the insect should be placed in a specimen tube filled with 70% Isopropyl alcohol (sometimes called Iso-propanol or Propan-2-ol). Isopropyl alcohol is available from most pharmacies and costs approximately £6 for 500ml. Once you have preserved your specimen in alcohol, a small slip of paper should be inserted in the tube with details of the location, date and grid reference of where the specimen was collected, together with the collector's name. These details should be written in pencil, as the alcohol will make ink fade or run.

Collecting Equipment

GB Nets
Unit 3A, Restormel Industrial Estate, Liddicoat Road, Lostwithiel, Cornwall PL22 0HG.
Telephone/Fax: 01208 873945
Email: sales@efe-uk.com   Website: www.gbnets-uk.com

Watkins and Doncaster
PO Box 5, Cranbrook, Kent TN18 5EZ. Telephone: 01580 753133. Fax: 01580 754054
Email: sales@watdon.com   Website: www.watdon.com

Alana Ecology Ltd
New Street, Bishop's Castle, Shropshire SY9 5DQ.
Telephone: 01588 630173. Fax: 01588 630176
Email: info@alanaecology.com   Website: www.alanaecology.com

THE EPHEMEROPTERA RECORDING SCHEME

The Ephemeroptera Recording Scheme would be grateful to receive records of any mayflies from throughout the British Isles. Further information can also be provided on the identification of the British Ephemeroptera. The scheme organiser is willing to confirm any identification you have made, but please contact the organiser before sending any voucher specimens for identification, as unsolicited specimens will be returned.
www.ephemeroptera.org.uk

Checklist of the British Ephemeroptera

This checklist details the species of Ephemeroptera known from the British Isles* and contains the most recent changes in nomenclature. It is based on the online Fauna Europaea checklist (www.faunaeur.org, 2009). Synonyms for recent changes are given.

AMELETIDAE	AMELETUS Bengtsson, 1865 *inopinatus* Eaton, 1887
ARTHROPLEIDAE	ARTHROPLEA Bengtsson, 1908 *congener* Bengtsson, 1908

BAETIDAE	BAETIS Leach, 1815
	atrebatinus Eaton, 1870
	    Labiobaetis atrebatinus (Eaton, 1870)
	buceratus Eaton, 1870
	digitatus (Bengtsson, 1912)
	fuscatus (Linnaeus, 1761)
	muticus (Linnaeus, 1758)
	    Alainites muticus (Linnaeus, 1758)
	niger (Linnaeus, 1761)
	    Nigrobaetis niger (Linnaeus, 1761)
	rhodani (Pictet, 1843)
	scambus Eaton, 1870
	vernus Curtis, 1834
	CENTROPTILUM Eaton, 1869
	luteolum (Müller, 1776)
	CLOEON Leach, 1815
	dipterum (Linnaeus, 1761)
	simile Eaton, 1870
	PROCLOEON Bengtsson, 1915
	bifidum (Bengtsson, 1912)
	pennulatum (Eaton, 1870)
	    Centroptilum pennulatum (Eaton, 1870)

CAENIDAE	BRACHYCERCUS Curtis, 1834 *harrisellus* Curtis, 1834 CAENIS Stephens, 1835 *beskidensis* Sowa, 1973 *horaria* (Linnaeus, 1758) *luctuosa* (Bürmeister, 1839) *macrura* Stephens, 1836 *pseudorivulorum* Keffermüller, 1960 *pusilla* Navás, 1913 *rivulorum* Eaton, 1884 *robusta* Eaton, 1884
EPHEMERIDAE	EPHEMERA Linnaeus, 1758 *danica* Müller, 1764 *lineata* Eaton, 1870 *vulgata* Linnaeus, 1758

EPHEMERELLIDAE EPHEMERELLA Walsh, 1862
              notata Eaton, 1887
          SERRATELLA Edmunds, 1959
              ignita (Poda, 1761)
                  Ephemerella ignita (Poda, 1761)

HEPTAGENIIDAE   ECDYONURUS Eaton, 1868
              dispar (Curtis, 1834)
              insignis (Eaton, 1870)
              torrentis Kimmins, 1942
              venosus (Fabricius, 1775)
          ELECTROGENA
              affinis (Eaton, 1883)
              lateralis (Curtis, 1834)
                  Heptagenia lateralis (Curtis, 1834)
          HEPTAGENIA Walsh, 1863
              longicauda (Stephens, 1836)
              sulphurea (Müller, 1776)
          KAGERONIA Matsumura, 1931
              fuscogrisea (Retzius, 1783)
                  Heptagenia fuscogrisea (Retzius, 1783)
          RHITHROGENA Eaton, 1881
              germanica Eaton, 1885
                  Rhithrogena haarupi Esbe-Petersen, 1909
              semicolorata (Curtis, 1834)

LEPTOPHLEBIIDAE HABROPHLEBIA Eaton, 1881
              fusca (Curtis, 1834)
          LEPTOPHLEBIA Westwood, 1840
              marginata (Linnaeus, 1767)
              vespertina (Linnaeus, 1758)
          PARALEPTOPHLEBIA Lestage, 1917
              cincta (Retzius, 1783)
              submarginata (Stephens, 1836)
              werneri Ulmer, 1919
                  Paraleptophlebia tumida Bengtsson, 1930

POTAMANTHIDAE   POTAMANTHUS Pictet, 1845
              luteus (Linneaus, 1767)

SIPHLONURIDAE   SIPHLONURUS Eaton, 1868
              alternatus (Say, 1824)
                  Siphlonurus linneanus Eaton, 1871
              armatus Eaton, 1870
              lacustris Eaton, 1870

* *Arthroplea congener* and *Heptagenia longicauda* may no longer occur in British waters.

Using this Guide

This guide provides a series of identification charts (ID charts) to both nymphs and adults. Identification charts 1A (for nymphs, page 9) and 1B (for adults, page 11) are designed to lead you to the correct Ephemeroptera family. On these charts you will find further details of the families and navigation commands leading to the next appropriate identification chart. Each family has its own identification chart and these are designed to lead you to a species identification. Once you have made a preliminary identification using the images you should refer to the species notes, which give further details, and also list similar species that may be misidentified. In all cases, you should consult a specialist key to confirm the identification (see page 126).

Nymphs

- Live underwater
- Three pairs of jointed legs
- Three relatively long tails (beware of breakages)
- Rounded head
- Abdominal gills present (usually attached to the sides, although in three families they are on the back, and in one family they are partially hidden by a pair of gill covers)

Go to Identification Chart 1A Page 9

Adults

- Found on the water surface, in bankside vegetation or flying close to water
- Three pairs of jointed legs
- Two or three long thread-like tails (beware of breakages)
- Wings membranous, with complex wing venation and sometimes distinctive patterning
- Wings held above the body (butterfly-like)

Go to Identification Chart 1B Page 11

Please note that the distribution maps provide an indication of the species distribution rather than an accurate record. Up to date distribution information can be obtained from the recording scheme organiser or the NBN Gateway (www.searchnbn.net). A calendar strip is included which indicates the months that the adults of each species are expected to emerge. It should be noted that local variations and weather conditions might affect the emergence of some species.

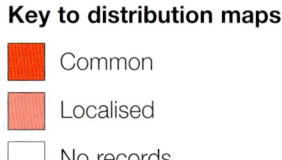

Key to distribution maps

- Common
- Localised
- No records

A Pictorial Guide to British Ephemeroptera 9

Identification Chart 1A: **Ephemeroptera Families – Nymphs**

1. Are the gills inserted at the sides of the body?

Yes → Go to 2

No → Go to 3
gills on back

2. Gills inserted at the side of the body
Compare your specimen to the three descriptions below

2a.
- Body flattened
- Rear edges of body segments drawn out into spines
- Large eyes on back of head*
- 7 pairs of plate-like gills, some of which also have a tuft of filaments
- Mature size: 8.5-14.0mm
- 5 free tarsal segments
- Antennae relatively short

*Warning: mature Baetidae can have large eyes on the back of the head, however they also have much smaller eyes at the side of the head.

Heptageniidae
(incl. Arthropleidae)

Nymphs are poor swimmers and prefer to cling to stones and boulders on lake shores and in streams and rivers

12 species
Go to ID Chart 6A
p. 74

2b.
- Body streamlined
- Rear edges of body segments rounded
- Small eyes on side of head*
- 6 or 7 pairs of plate-like gills, some of which may be double
- Mature size: 4.5-11.0mm
- 3 free tarsal segments
- Antennae relatively long

*Warning: mature Baetidae can also have large eyes on the back of the head.

Baetidae

Nymphs are good swimmers and are found in streams, rivers, ponds and lakes

14 species
Go to ID Chart 2A
p. 13

2c.
- Body streamlined
- Rear edges of body segments drawn out into spines*
- Small eyes on back of head
- 7 pairs of plate-like gills, some of which may be double
- Mature size: 8.5-18.0mm
- 4 free tarsal segments
- Antennae relatively short

*Warning: in the Ameletidae the spines are very small

Siphlonuridae
(incl. Ameletidae)

Nymphs are good swimmers and are found in rivers, ponds and lakes

4 species
Go to ID Chart 8A
p. 116

3. Are the gills branched or simple?

Branched → Go to 4

Simple → Go to 5

4. Branched gills

Compare your specimen to the three descriptions below

4a.
- Body streamlined
- Gills feathery
- Gills held at side of body
- Tails shorter than the body
- Mature size: 8.0-15.0mm

Nymphs found chiefly in large rivers, often in side pools with a bottom of stone and sand

Potamanthidae

1 species
Potamanthus luteus
p. 115

4b.
- Body cylindrical
- Gills feathery
- Gills held over back
- Tails much shorter than the body
- Mature size: 18.0-25.0mm

Nymphs burrow in gravel, sand or mud in still or flowing water

Ephemeridae

3 species
Go to ID Chart 4A
p. 60

4c.
- Body streamlined
- Gills strap-like, plate-like with a terminal filament, or with many filaments
- Tails longer than body and held in a T shape at rest
- Mature size: 5.0-12.0mm

Nymphs crawl on the bed or swim in a laboured fashion. They are found in ponds, lakes, streams and rivers

Leptophlebiidae

6 species
Go to ID Chart 7A
p. 100

5. Simple gills

Compare your specimen to the two descriptions below

5a.
- Body squat
- Gills stacked under a large pair of gill covers
- Mature size: 5.0-12.0mm

Nymphs rarely swim and are usually found crawling in mud and debris in still and flowing water

Caenidae

9 species
Go to ID Chart 3A
p. 42

5b.
- Body streamlined
- 4 pairs of plate-like gills visible
- Mature size: 7.0-10.0mm

Nymphs swim poorly with a characteristic 'S' shaped motion. Usually found amongst stones or amongst mosses and plants in streams, rivers and occasionally lakes.

Ephemerellidae

2 species
Go to ID Chart 5A
p. 68

Identification Chart 1B: Ephemeroptera Families – Adults

1. Does your specimen have three tails?

Make sure that you count any broken tails. The easiest way of determining whether your specimen has two or three tails is to look where the middle tail should be. In species with two tails there is a short rounded stump in place of the middle tail. In species with three tails, where the middle tail is broken the stump is open.

three tails only — **Yes** Go to 2 | **No** Go to 3 — two tails only

2. Three tails

Check the wings and compare your specimen to the descriptions below

2a.
- Large fly
- Dark marks on forewings
- Forewing size: 15.0-25.0mm

Adults are large flies with three tails and large hindwings

Ephemeridae — 4 species Go to ID Chart 4B p. 61

2b.
- Medium to large sized fly
- Yellow in colour
- Forewing size: 8.5-18.0mm

Adults are medium to large flies with three tails and large hindwings

Potamanthidae — 1 species *Potamanthus luteus* p. 115

2c.
- Small to medium sized fly
- Regular short veins between each long vein in forewing
- Forewing size: 7.0-11.0mm

Adults are small to medium sized flies with three tails and large hindwings

Ephemerellidae — 2 species Go to ID Chart 5B p. 69

2d.
- Small to medium sized fly
- No regular short veins on forewing
- Forewing size: 6.0-13.0mm

Adults are small to medium sized flies with three tails and large hindwings

Leptophlebiidae — 6 species Go to ID Chart 7B p. 101

2e.
- Very small fly
- No hindwings present
- Forewing size: 3.0-9.0mm

Adults are very small flies with three tails and no hindwings

Caenidae — 9 species Go to ID Chart 3B p. 43

3. Two tails

Check the hindwings and compare your specimen to the descriptions below

3a.
- Medium to large sized fly
- Large hindwings
- Four free segments in the foot
- Forewing size: 8.0-16.0mm

Adults are medium to large flies with two tails and large hindwings

Siphlonuridae
(incl. Ameletidae)

4 species
Go to ID Chart 8B
p. 117

3b.
- Medium to large sized fly
- Large hindwings
- Five free segments in the foot
- Forewing size: 8.0-17.0mm

Adults are medium to large sized flies with two tails and large hindwings

Heptageniidae

12 species
Go to ID Chart 6B
p. 75

3c.
- Small to medium sized fly
- Small hindwings, absent in some species
- Three free segments in the foot
- Forewing size: 5.0-12.0mm

Adults are small flies with two tails and small or apparently absent hindwings. In three species the hindwings are absent

Baetidae

4 species
Go to ID Chart 2B
p. 14

Identification Chart 2A: Baetidae – Nymphs

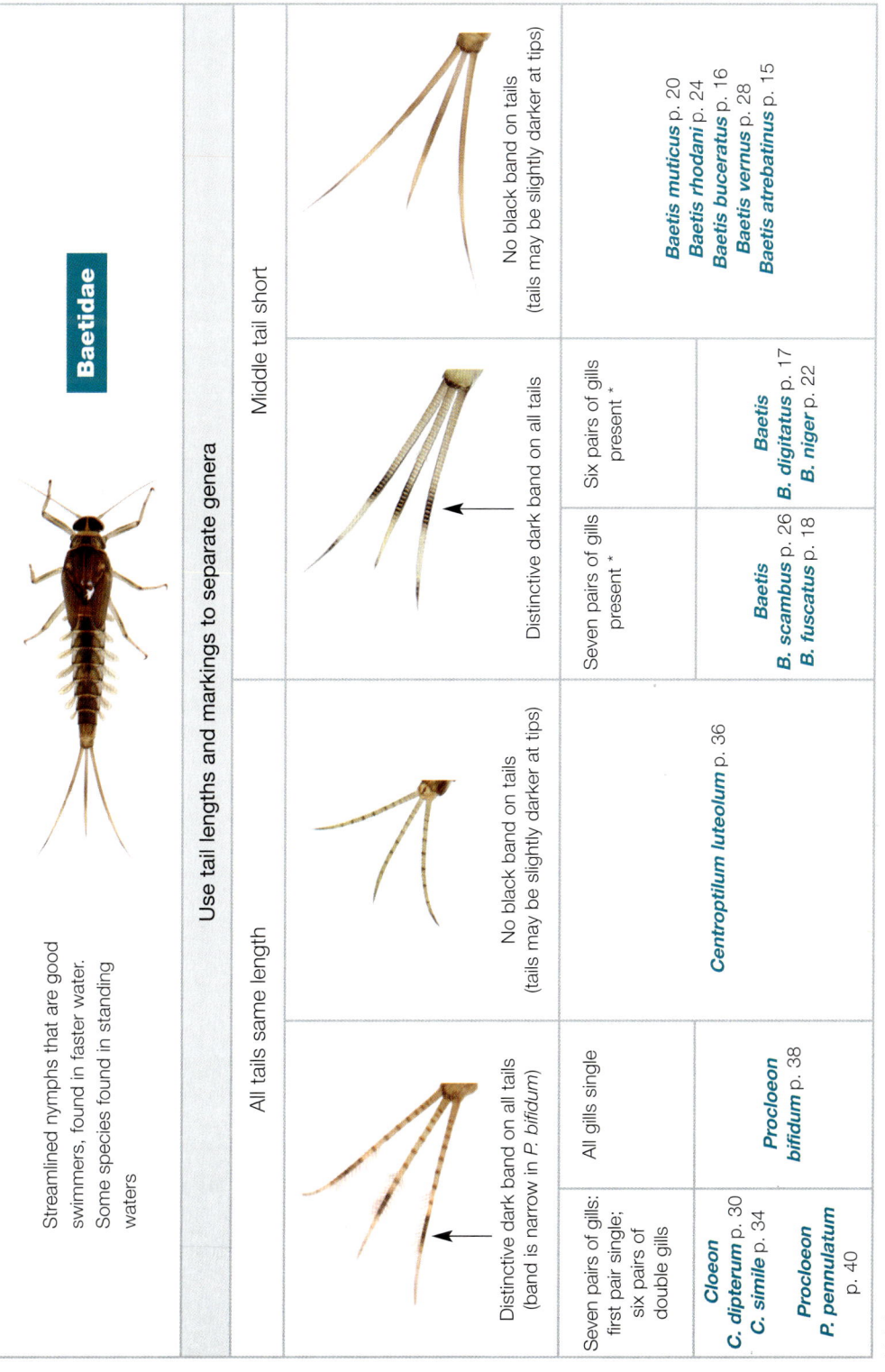

Baetidae

Streamlined nymphs that are good swimmers, found in faster water. Some species found in standing waters

Use tail lengths and markings to separate genera

	Middle tail short			
	No black band on tails (tails may be slightly darker at tips)			**Baetis muticus** p. 20 **Baetis rhodani** p. 24 **Baetis buceratus** p. 16 **Baetis vernus** p. 28 **Baetis atrebatinus** p. 15
	Distinctive dark band on all tails	Six pairs of gills present *		
		Seven pairs of gills present *	**Baetis** **B. scambus** p. 26 **B. fuscatus** p. 18	**Baetis** **B. digitatus** p. 17 **B. niger** p. 22

	All tails same length		
	No black band on tails (tails may be slightly darker at tips)		**Centroptilum luteolum** p. 36
	Distinctive dark band on all tails (band is narrow in *P. bifidum*)	All gills single	**Procloeon bifidum** p. 38
		Seven pairs of gills: first pair single; six pairs of double gills	**Cloeon** **C. dipterum** p. 30 **C. simile** p. 34 **Procloeon** **P. pennulatum** p. 40

* It is important to check the number of gills carefully. In damaged specimens the last (7th) pair of gills is often missing resulting in six pairs of gills being present. In *Baetis digitatus* and *B. niger*, it is always the first gill that is absent. In mature nymphs the 1st gill is sometimes hidden under the developing wing pads, or in the case of *B. muticus* and *B. atrebatinus* the 1st gill is very small.

Identification Chart 2B: Baetidae – Adults

Baetidae

Small to medium sized flies with two tails and small hindwings (absent in some species).
Forewing size: 5.0–12.0 mm

Use shape and venation to separate genera

Hindwings present

Oval hindwing with forked veins	*Baetis muticus* p. 20 *Baetis digitatus* p. 17 *Baetis niger* p. 22
Oval hindwings – all veins single; Top edge of hindwing with small pointed projection	*B. buceratus* p. 16 *B. fuscatus* p. 18 *B. rhodani* p. 24 *B. scambus* p. 26 *B. vernus* p. 28
Oval hindwings – all veins single; Top edge of hindwing smooth	*Baetis atrebatinus* p. 15
Spur-shaped hindwings; Hindwing with rounded tip	*Procloeon pennulatum* p. 40
Spur-shaped hindwings; Hindwing with pointed tip	*Centroptilum luteolum* p. 36

Hindwings absent

First segment of foot is about twice the length of the second	*Cloeon* *C. dipterum* p. 30 *C. simile* p. 34
First segment of foot is about three times the length of the second	*Procloeon bifidum* p. 38

Baetis atrebatinus Eaton
Common name: Dark Olive

Key features

Nymphs: Streamlined nymphs with 7 pairs of plate-like gills, with the first pair much smaller than the others.

Adults: Small to medium sized flies with two tails and small oval hindwings. The sub-imago has pale grey wings with pale brown veins. Both the male and the female have dark olive-green bodies, although in the male the last segment is more yellowish in colour. The legs are also dark olive-green and they have much darker feet. The tails are dull grey. In the female the eyes are green, while in the male they are pale brick-red.

The imago has transparent wings with brown veins. The legs are dark olive-grey with darker feet and the tails are grey with reddish brown rings at each segment. The female imago has a reddish-brown body with pale olive rings defining the rear edge of the body segments. The eyes are a dark brown. The male imago has a pale olive body which has traces of brown in places. The last three segments are an orange-brown colour. In both sexes, the eyes of the imago are pale brownish-red.

Separating from other species

Nymphs: *Baetis atrebatinus* has narrow-bodied nymphs that superficially resemble those of *B. muticus*. They can however be separated by the presence of a small bump on the outer edge of the bottom segment of the antenna.

Adults: Adults of *Baetis atrebatinus* are superficially similar to those of *B. rhodani*, however they are slightly smaller and the costal process (a small projection on the leading edge) on the hindwing is absent in *B. atrebatinus*.

Habitat and ecology

Nymphs of this species typically live in the riffle areas of calcareous rivers and streams where they swim in short, darting bursts amongst the substrate. They feed by scraping algae from submerged stones and other structures, or by gathering or collecting fine particulate organic detritus from the sediment.

Very little is known about the life cycle of *B. atrebatinus*, however adults have been collected between May and October, which may suggest that there is more than one generation per year.

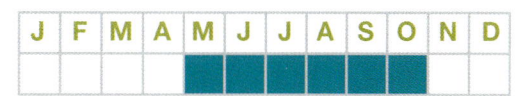

J	F	M	A	M	J	J	A	S	O	N	D
				■	■	■	■	■	■		

Emergence of the adults probably takes place at the surface of the water. However very little is known about the swarming behaviour of this species other than that adults are found flying throughout the day and into the evening.

There have been no studies on the egg-laying habits of *B. atrebatinus*. However it is likely that the female will lay around 1,200 eggs either directly on a partly submerged stone or by releasing her eggs in several batches onto the water surface.

Distribution

Baetis atrebatinus is a localised species with records from Hampshire and Dorset, Eire and the River Teifi in Wales. A voucher specimen is required for records from other areas.

Baetis buceratus Eaton
Common name: Scarce Olive

Key features

Nymphs: Streamlined nymphs with 7 pairs of plate-like gills, with the first and last pair of a similar size.

Adults: Small flies with two tails and small oval hindwings. The male sub-imago has dull grey wings, the veins of which are tinged with golden-olive. The body is dark grey to medium olive, with the underside of each segment more yellowish. The eyes are dull red-brown. The female sub-imago has dull grey wings with light brown venation and brown to medium olive bodies. Their eyes are dull yellow-green.

In the imago, both the male and female have transparent wings with light brown veins, grey-olive legs and long, off-white tails. The female has a dark reddish-brown body, whereas the male has a grey-olive body, with the last three segments reddish-brown. In both sexes the eyes are brown.

Separating from other species

Nymphs: *Baetis buceratus* is superficially similar to both *B. vernus* and *B. rhodani*. It can be readily separated from *B. rhodani* by the absence of spines amongst the hairs on the edge of the gills (compare with *B. rhodani*, p.24). *B. buceratus* can be separated from *B. vernus* by the presence of a series of 2 to 4 light dots on each body segment in *B. buceratus*.

Adults: Sub-imagines of *Baetis buceratus* and *B. vernus* are almost inseparable. Imagines can be separated by small differences in the male genitalia.

Habitat and ecology

Nymphs of this species typically live amongst the sand and gravel in the riffle areas of rivers where they swim in short, darting bursts amongst the substrate. They feed by scraping algae from submerged stones and other structures, or by gathering or collecting fine particulate organic detritus from the sediment.

Little is known about the life cycle of this species however adults have been collected between April and October. It is thought that there are two generations a year, one with overwintering eggs that emerge in the spring and another that grows over the summer and emerges later in the year. Emergence of the adults probably takes place at the surface of the water during daylight hours.

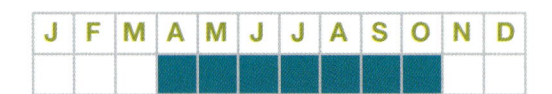

J	F	M	A	M	J	J	A	S	O	N	D
		■	■	■	■	■	■	■	■		

Distribution

Baetis buceratus is a widespread, though localised species. It has been recorded from rivers in Wales, the Midlands and the south of England, but is absent from Scotland, Northern England and Ireland, and as such a voucher specimen is required for records from these areas.

Baetis digitatus (Bengtsson)
Common name: Scarce Iron Blue

Key features

Nymphs: Streamlined nymphs with six pairs of plate-like gills, the first pair (nearest the head) are absent. Tails with a median black band.

Adults: Small flies with two tails and small oval hindwings. The sub-imago has dull yellow-green eyes in the female and dull red-brown eyes in the male. In both sexes the wings are a dull grey-blue colour and the body is dark brown, which is sometimes tinged with olive. The legs are pale to dark olive-brown, whilst the tails are dark grey. The imago has transparent wings and pale grey legs and tails. The body in the female is dark claret-brown, whilst in the male it is translucent white, with the last three segments being dark orange-brown.

Separating from other species

Nymphs: *Baetis digitatus* and *B. niger* are both narrow-bodied nymphs which have a black band across their tails. *Baetis digitatus* and *B. niger* are the only Baetidae with only 6 pairs of plate-like gills. Separation of nymphs of *B. digitatus* and *B. niger* can be achieved by examination of the last gill. In *B. digitatus* the hind edge of the gill is slightly concave resulting in a different shape when compared to other Baetidae species.

Adults: *B. digitatus* and *B. niger* can be distinguished from other adult Baetidae by the presence of only two veins on the oval-shaped hindwing, the second of which is forked. The only other Baetidae species with a forked vein is *Baetis muticus*, however this species has a third vein which runs along the lower edge of the wing. It is often difficult to see the third vein, and care must be taken to double check whether it is present or not. Separation of adults of *B. digitatus* and *B. niger* is not reliable at present.

Habitat and ecology

Nymphs of this species typically crawl amongst in-stream vegetation in riffle areas of rivers and streams or swim in short, darting bursts amongst the substrate. They feed by scraping algae from submerged stones and other structures, or by gathering or collecting fine particulate organic detritus from the sediment.

Little is known about the life cycle of this species however adults have been found between May and September. Recent work has shown that there appears to be two distinct peaks in the flight period – one in the spring and another in the autumn. This may suggest that there are two generations per year – a slow growing winter generation and a much faster summer generation.

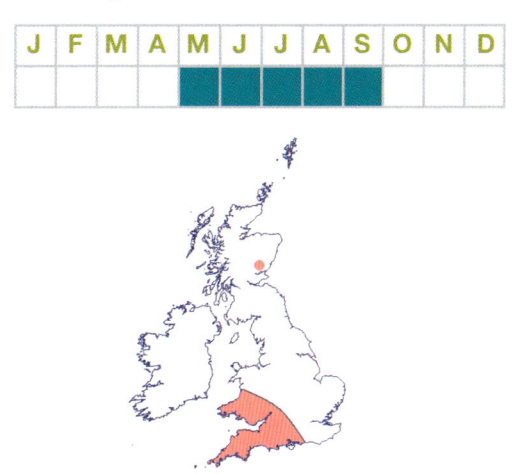

J	F	M	A	M	J	J	A	S	O	N	D
				■	■	■	■	■			

Distribution

B. digitatus is a highly localised species with records from a small number of watercourses in Wales, southern England and a single watercourse in Scotland. A voucher specimen is required for all records.

Baetis fuscatus (Linnaeus)
Common name: Pale Watery

Life size

Male sub-imago

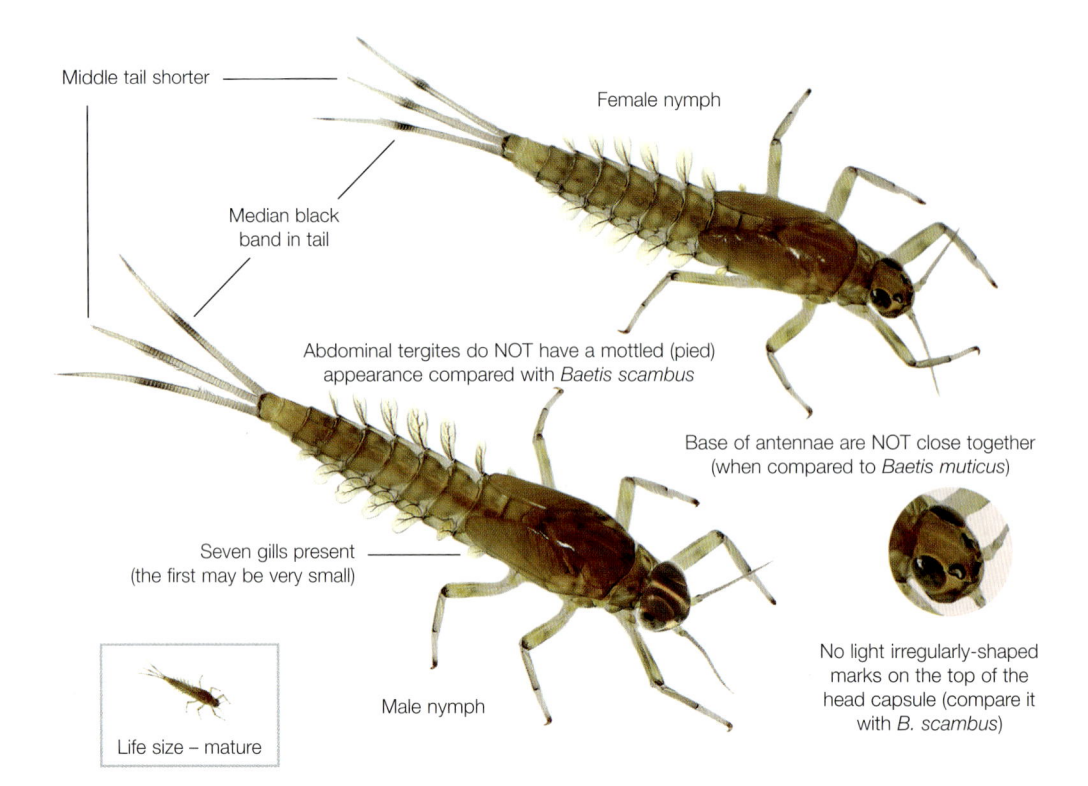

Middle tail shorter

Female nymph

Median black band in tail

Abdominal tergites do NOT have a mottled (pied) appearance compared with *Baetis scambus*

Base of antennae are NOT close together (when compared to *Baetis muticus*)

Seven gills present (the first may be very small)

Male nymph

Life size – mature

No light irregularly-shaped marks on the top of the head capsule (compare it with *B. scambus*)

Baetis fuscatus (Linnaeus)
Common name: Pale Watery

Key features

Nymphs: Streamlined nymphs with 7 pairs of plate-like gills, with the first and last pair of a similar size. Tails with a median black band.

Adults: Small flies with two tails and small oval hindwings. The sub-imago has pale grey wings, which have pale brown veins. They have distinctive pale olive bodies, the last two segments of which are pale yellow. The legs are pale olive, with dark grey feet, whilst the tails are a uniform grey colour. The eyes of the female are yellow-green, whilst those of the male are bright orange-yellow.

The male imago has distinctive lemon-yellow eyes however the body, legs and tails are predominately translucent creamy-white with slight touches of olive. The last three segments of the body are orange-brown. In the female the body is a distinctive golden-olive colour while the legs are pale olive-brown and the eyes are dark brown. The last three segments of the body are orange-brown in the male and dark golden-olive in the female.

Separating from other species

Nymphs: *Baetis fuscatus* is almost inseparable from the nymphs of *Baetis scambus*. They are both wide-bodied nymphs with a dark band across their tails. Some workers have used irregular shaped marks on the head to separate *B. fuscatus* from *B. scambus* however this feature is unreliable.

Adults: Adults of both species are similarly difficult to separate. Females and sub-imagines are inseparable at present. Males of *B. fuscatus* can be separated from *B. scambus* by their distinctive eyes, however they are morphologically similar leading to some workers suggesting that these two species are in fact merely different races of the same species.

Habitat and ecology

Nymphs of this species typically live in the riffle areas of rivers and streams either on in-stream vegetation or amongst the sand and gravel on the bed. The nymphs swim in short, darting bursts amongst the substrate or climb amongst the vegetation. They feed by scraping algae from submerged stones and other structures, or by gathering or collecting fine particulate organic detritus from the sediment. There is generally one generation per year that overwinters in the egg stage, with adults present between May and October. Some workers have suggested that there may be two or more generations per year.

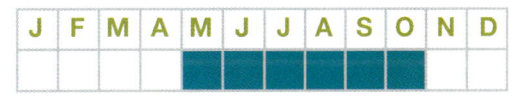

J	F	M	A	M	J	J	A	S	O	N	D
				■	■	■	■	■			

Emergence of the adults probably takes place at the surface of the water during daylight hours. The males of this species swarm at dusk, but may also swarm at other times of the day.

Once mated, the female flies to the river and lands on a partly submerged stone. She then folds her wings and pulls herself under the water to find a suitable place to lay around 1,200 eggs. The eggs are laid individually alongside each other to form a contiguous patch of eggs. Once completed, she will sometimes climb back out of the water and fly away. However more often than not, she will be swept away by the current.

Distribution

Due to the problems with the identification of nymphs, it is difficult to compile a complete distribution for this species. *Baetis fuscatus* is, however, thought to be less common than *Baetis scambus*.

Baetis muticus **(Linnaeus)**
Common name: Iron Blue

Life size

Male sub-imago

Middle tail much shorter than outer tails

No distinct black band near the middle of the tail

Seven gills present 1st much smaller than 7th

Head of female

Pillared eyes = male

Base of antennae close together

Basal segment of antenna symmetrical, without lobe on one side

Life size – mature

Nymph

Baetis muticus (Linnaeus)
Common name: Iron Blue

Key features

Nymphs: Streamlined nymphs with 7 pairs of plate-like gills, with the first pair much smaller than the others.

Adults: Small flies with two tails and small oval hindwings. The sub-imago has dull grey-blue wings and a dark brown body that is sometimes tinged with olive. The legs are pale to dark olive-brown, whilst the tails are dark grey. The imago has transparent wings and pale grey legs and tails. The body in the female is dark claret-brown, whilst in the male it is translucent white, with the last three segments being dark orange-brown.

Separating from other species

Nymphs: *Baetis muticus* are distinctive narrow-bodied nymphs with the antennae inserted closely together between the eyes. They do not have a distinct black band across the tails. There are seven pairs of gills, although the first pair is very small and easily overlooked in mature nymphs.

Adults: *Baetis muticus* can be distinguished from other adult Baetidae by the presence of three veins on the oval hind wing, the second of which is forked. The third vein, which runs along the lower edge of the wing, is missing in *B. digitatus* and *B. niger*. It is often difficult to see the third vein, and care must be taken to double check whether it is present or not.

Habitat and Ecology

Nymphs of this species live chiefly in riffle sections of rivers and streams, where they are found in gravel, sand or mud on the bed of the watercourse. The nymphs are good swimmers and typically swim in short, darting bursts. They feed by scraping algae from submerged stones and other structures, or by gathering or collecting fine particulate organic detritus from the sediment.

There are two generations per year – a slow growing winter generation and a much faster summer generation. This results in a fairly long flight period, with adults being present between April and October.

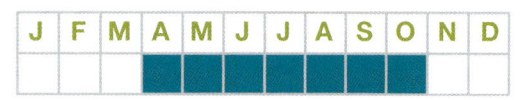

J	F	M	A	M	J	J	A	S	O	N	D
			■	■	■	■	■	■	■		

Emergence of the adults typically takes place on the surface of the water during daylight hours. The males of this species can be found swarming throughout the day, however swarming stops at the onset of dusk.

Once mated, the female flies to the river and lands on a partly submerged stone. She then folds her wings and pulls herself under the water to find a suitable place to lay around 3,500 eggs. The eggs are laid individually alongside each other to form a contiguous patch of eggs. Once completed, she will sometimes climb back out of the water and fly away. However more often than not, she will be swept away by the current. In some cases the female will fly to the river, where she descends to the surface of the water and releases her eggs in several batches by dipping the tip of her abdomen into the water surface.

Distribution

Baetis muticus is a widespread and common species that is found throughout the British Isles. It is relatively uncommon in the south-east of England.

Baetis niger (Linnaeus)
Common name: Southern Iron Blue

Life size

Male sub-imago

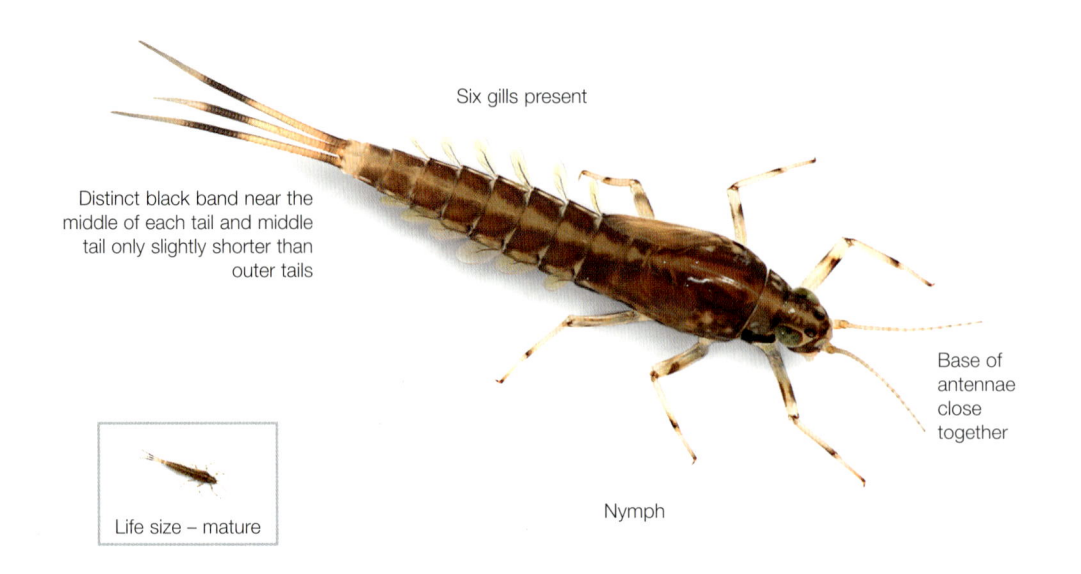

Six gills present

Distinct black band near the middle of each tail and middle tail only slightly shorter than outer tails

Base of antennae close together

Life size – mature

Nymph

Baetis niger (Linnaeus)
Common name: Southern Iron Blue

Key features

Nymphs: Streamlined nymphs with 6 pairs of plate-like gills, the first pair (nearest the head) are absent. Tails with a median black band.

Adults: Small flies with two tails and small oval hindwings. The sub-imago has dull yellow-green eyes in the female and dull red-brown eyes in the male. In both sexes the wings are a dull grey-blue colour and the body is dark brown which is sometimes tinged with olive. The legs are pale to dark olive-brown, whilst the tails are dark grey. The imago has transparent wings and pale grey legs and tails. The body in the female is dark claret-brown, whilst in the male it is translucent white, with the last three segments being dark orange-brown.

Separating from other species

Nymphs: *Baetis digitatus* and *B. niger* are both narrow-bodied nymphs which have a black band across their tails. In addition to the central dark band on the tails *B. niger* has a faint dark band at the base of the tails. *Baetis niger* and *B. digitatus* are the only Baetidae with only 6 pairs of plate-like gills. Separation of nymphs of *B. digitatus* and *B. niger* can be achieved by examination of the last gill. In *B. digitatus* the hind edge of the gill is slightly concave resulting in a different shape when compared to other Baetidae species, including *N. niger*.

Adults: *Baetis niger* and *B. digitatus* can be distinguished from other adult Baetidae by the presence of only two veins on the oval-shaped hind wing, the second of which is forked. The only other Baetidae species with a forked vein is *Baetis muticus*, however this species has a third vein which runs along the lower edge of the wing. It is often difficult to see the third vein, and care must be taken to double check whether it is present or not. Separation of adults of *B. niger* and *B. digitatus* is not reliable at present.

Habitat and Ecology

Nymphs of this species typically crawl amongst in-stream vegetation in riffle areas of rivers and streams or swim in short, darting bursts amongst the substrate. They feed by scraping algae from submerged stones and other structures, or by gathering or collecting fine particulate organic detritus from the sediment.

There are two generations per year – a slow growing winter generation and a much faster summer generation. This results in a fairly long flight period, with adults being present between April and October. Recent work on the River Test has found that the summer generation is considerably reduced, if not absent.

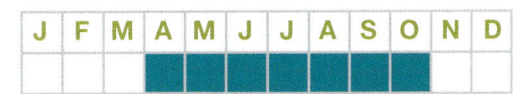

J	F	M	A	M	J	J	A	S	O	N	D
			■	■	■	■	■	■	■		

Emergence of the adults is thought to occur at the surface of the water during daylight hours. The males of this species can be found swarming throughout the day, and often swarming continues until dusk.

Once mated, the female fly either pulls herself under the water surface to lay around 1,200 eggs directly on a partly submerged stone. In some cases she will fly to the river, where she descends to the surface of the water and releases her eggs in several batches by dipping the tip of her abdomen onto the water surface.

Distribution

Baetis niger is a widespread, though localised species found in running waters throughout the British Isles with the exception of Ireland. As a result, a voucher specimen is required for records from Ireland.

Baetis rhodani (Pictet)
Common name: Large Dark Olive

Life size

Female sub-imago

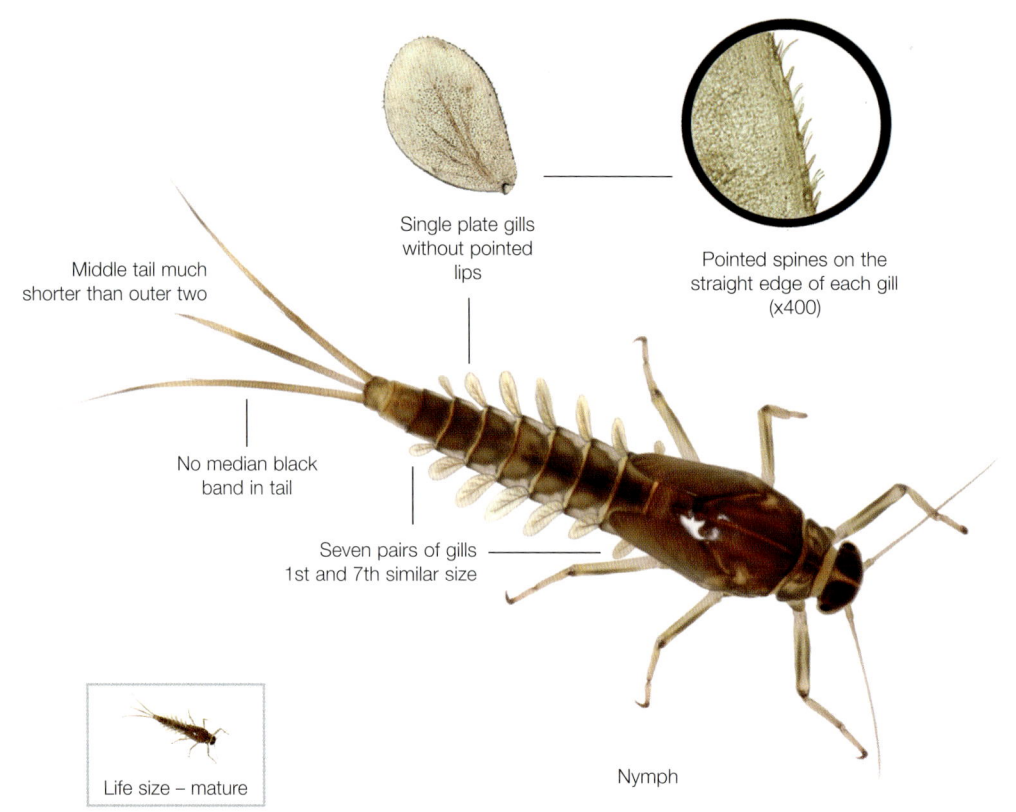

Middle tail much shorter than outer two

Single plate gills without pointed lips

Pointed spines on the straight edge of each gill (x400)

No median black band in tail

Seven pairs of gills 1st and 7th similar size

Life size – mature

Nymph

Baetis rhodani (Pictet)
Common name: Large Dark Olive

Key features

Nymphs: Streamlined nymphs with 7 pairs of plate-like gills, with the first and last pair of a similar size.

Adults: Small to medium sized flies with two tails and small oval hindwings. The sub-imago has pale grey wings with pale brown veins. Both the male and the female have dark olive-green bodies, although in the male the last segment is more yellowish in colour. The legs are also dark olive-green and they have much darker feet. The tails are dull grey. In the female the eyes are green, while in the male they are dull brick-red.

The imago has transparent wings with brown veins. The legs are dark olive-grey with darker feet and the tails are grey with reddish-brown rings at each segment. The female imago has a reddish-brown body with pale olive rings defining the rear edge of the body segments. The eyes are a dark brown. The male imago has a pale olive body that has traces of brown in places. The last three segments are an orange-brown colour. In both sexes, the eyes of the imago are dark red.

Separating from other species

Nymphs: *Baetis rhodani* is readily separated from all other *Baetis* species by the presence of small spines amongst the hairs along the edge of the gills (see opposite).

Adults: *Baetis rhodani* is the largest *Baetis* species found in Britain. Sub-imagines of *B. rhodani* can be separated from other species by the presence of two small open circles on the upper side of the body segment to which the hindwings are attached (see below).

Habitat and Ecology

Nymphs of this species typically live in the riffle areas of rivers and streams amongst the sand and gravel on the bed where they swim in short, darting bursts amongst the substrate. They feed by scraping algae from submerged stones and other structures, or by gathering or collecting fine particulate organic detritus from the sediment.

There are typically two generations a year, one with overwintering nymphs that emerge in the spring and another that grows over the summer and emerges later in the year. In warmer years there may be further generations, resulting in adults of this species being recorded in every month.

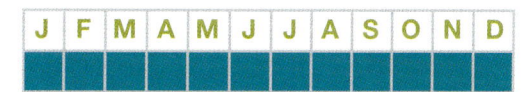

J	F	M	A	M	J	J	A	S	O	N	D
■	■	■	■	■	■	■	■	■	■	■	■

Emergence of the adults takes place at the surface of the water during daylight hours and at dusk. The males of this species swarm throughout the day but typically swarming stops before dusk.

Once mated, the female flies to the river and lands on a partly submerged stone. She then folds her wings and pulls herself under the water to find a suitable place to lay around 4,500 eggs. The eggs are laid individually alongside each other to form a contiguous patch of eggs. Once completed, she will sometimes climb back out of the water and fly away. However more often than not, she will be swept away by the current.

Distribution

Baetis rhodani is one of the most common and widespread Ephemeroptera species. It can be found throughout the British Isles.

Baetis scambus Eaton
Common name: Small Dark Olive

Life size

Male sub-imago

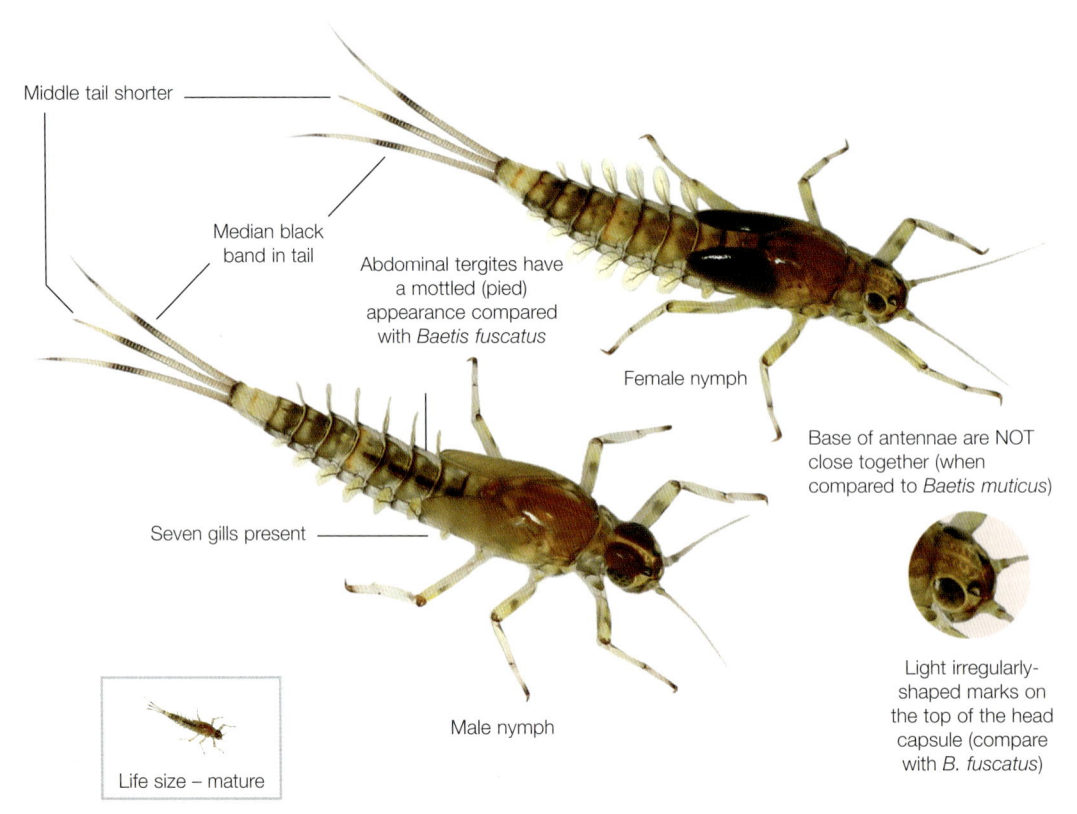

Middle tail shorter

Median black band in tail

Abdominal tergites have a mottled (pied) appearance compared with *Baetis fuscatus*

Female nymph

Base of antennae are NOT close together (when compared to *Baetis muticus*)

Seven gills present

Male nymph

Life size – mature

Light irregularly-shaped marks on the top of the head capsule (compare with *B. fuscatus*)

Baetis scambus Eaton
Common name: Small Dark Olive

Key features

Nymphs: Streamlined nymphs with 7 pairs of plate-like gills, with the first and last pair of a similar size. Tails with a median black band.

Adults: Small flies with two tails and small oval hindwings. The sub-imago has a medium to dark grey to grey-olive body, the last two segments of which are yellowish. The legs are pale olive, with almost black feet, whilst the tails are a uniform grey colour. The joints of the legs become progressively lighter as the flight period continues. The eyes of the female are almost black, whilst those of the male are dull orange-red.

The imago has olive-brown legs and greyish-white tails. In both the male and the female the wings are transparent, however in the female the veins are much darker. In the male the body is a translucent cream colour with the last three segments being an opaque orange-brown colour. The colour of the body in the female varies between pale brown to deep reddish-brown with age. The eyes in the male are bright orange-red, while the female has almost black eyes.

Separating from other species

Nymphs: *Baetis scambus* is almost inseparable from the nymphs of *Baetis fuscatus*. They are both wide-bodied nymphs with a dark band across their tails. Some workers have used irregularly shaped marks on the head of *B. fuscatus* to separate it from *B. scambus*, however this feature is unreliable.

Adults: Adults of both species are similarly difficult to separate. Females and sub-imagines are inseparable at present. Males of *B. fuscatus* can be separated from *B. scambus* by their distinctive eyes, however they are morphologically similar leading to some workers suggesting that these two species are in fact merely different races of the same species.

Habitat and Ecology

Nymphs of this species typically live in the riffle areas of rivers and streams either on in-stream vegetation or amongst the sand and gravel on the bed. The nymphs swim in short, darting bursts amongst the substrate or climb amongst the vegetation. They feed by scraping algae from submerged stones and other structures, or by gathering or collecting fine particulate organic detritus from the sediment.

There are two generations a year, one with overwintering eggs that emerges in the spring and another that grows over the summer and emerges later in the year. This results in a long flight period with adults present between February and November.

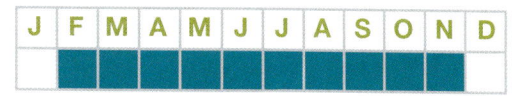

J	F	M	A	M	J	J	A	S	O	N	D
	■	■	■	■	■	■	■	■	■	■	

Emergence of the adults probably takes place at the surface of the water during daylight hours. The males of this species swarm throughout the day but typically stop before dusk.

Once mated, the female flies to the river and lands on a partly submerged stone. She then folds her wings and pulls herself under the water to find a suitable place to lay around 1,200 eggs. The eggs are laid individually alongside each other to form a contiguous patch of eggs. Once completed, she will sometimes climb back out of the water and fly away. However more often than not, she will be swept away by the current.

Distribution

Due to the problems with the identification of nymphs, it is difficult to compile a complete distribution for this species. *Baetis scambus* is, however, thought to be more common than *Baetis fuscatus*.

Baetis vernus **Curtis**
Common name: Medium Olive

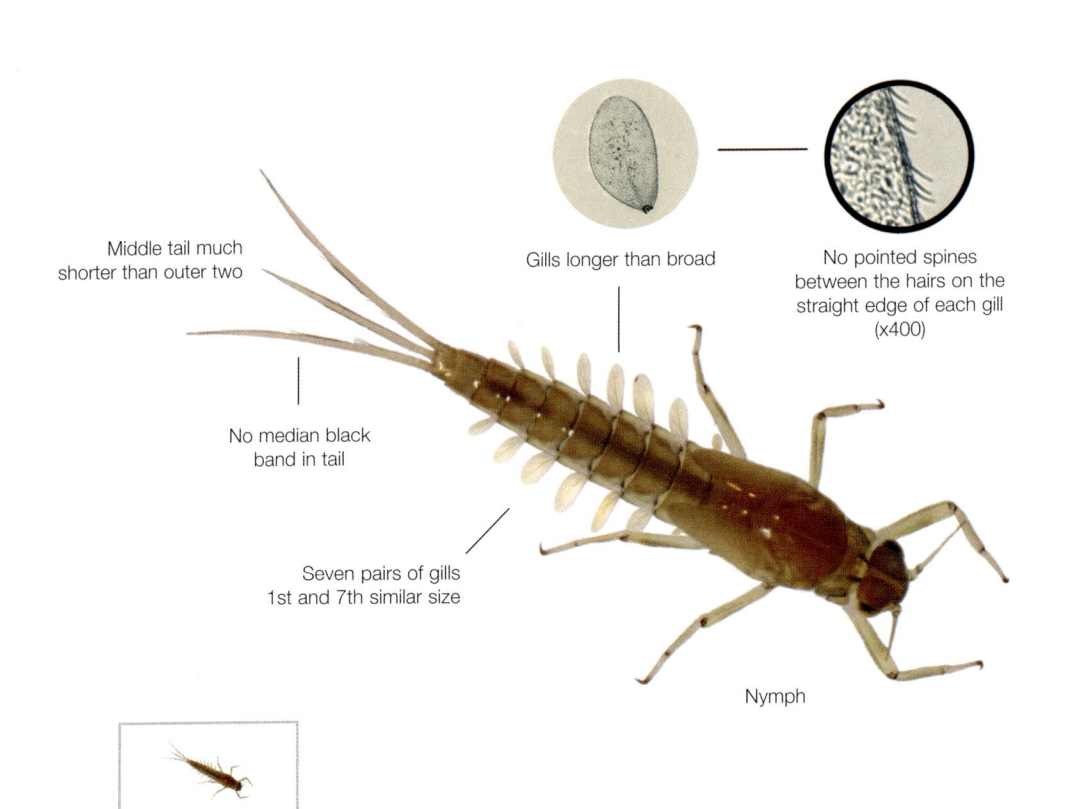

Life size

Female sub-imago

Middle tail much shorter than outer two

Gills longer than broad

No pointed spines between the hairs on the straight edge of each gill (x400)

No median black band in tail

Seven pairs of gills 1st and 7th similar size

Nymph

Life size – mature

Baetis vernus Curtis
Common name: Medium Olive

Key features

Nymphs: Streamlined nymphs with 7 pairs of plate-like gills, with the first and last pair of a similar size.

Adults: Small to medium sized flies with two tails and small oval hindwings. The male sub-imago has dull grey wings, the veins of which are tinged with golden olive. The body is dark grey to medium olive, with the underside of each segment more yellowish. The eyes are dull red brown. The female sub-imago has dull grey wings with light brown venation and a brown to medium olive body. The eyes are dull yellow-green.

In the imago, both the male and female have transparent wings with light brown veins, grey olive legs and long, off-white tails. The female has a dark reddish-brown body, whereas the male has a grey-olive body, with the last three segments reddish-brown. In both sexes the eyes are brown.

Separating from other species

Nymphs: *Baetis vernus* is superficially similar to both *B. buceratus* and *B. rhodani*. It can be readily separated from *B. rhodani* by the absence of spines amongst the hairs on the edge of the gills (compare with *B. rhodani*, p. 24). *B. vernus* can be separated from *B. buceratus* by the absence of a series of 2 to 4 light dots on each body segment (compare with *B. buceratus*, p. 16).

Adults: Sub-imagines of *Baetis vernus* and *B. buceratus* are almost inseparable. Imagines can be separated by small differences in the male genitalia.

Habitat and Ecology

Nymphs of this species live chiefly in the pools and margins of rivers and streams, where they live amongst the sand and gravel of the bed or on in-stream vegetation. The nymphs swim in short, darting bursts amongst the substrate or climb amongst the vegetation. They feed by scraping algae from submerged stones and other structures, or by gathering or collecting fine particulate organic detritus from the sediment.

There are two generations a year, one with overwintering eggs that emerges in the spring and another that grows over the summer and emerges later in the year. In some years there may be further generations, but adults are generally only seen between April and October.

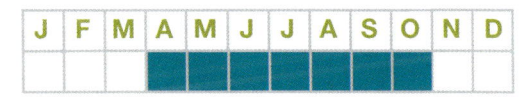

J	F	M	A	M	J	J	A	S	O	N	D
			■	■	■	■	■	■	■		

Emergence of the adults takes place at the surface of the water during the day and at dusk. The males swarm throughout the day but typically stop before dusk.

Once mated, the female flies to the river and lands on a partly submerged stone. She then folds her wings and pulls herself under the water to find a suitable place to lay her eggs. The eggs are laid individually alongside each other to form a contiguous patch of eggs. Once completed, she will sometimes climb back out of the water and fly away. However more often than not, she will be swept away by the current.

Distribution

Baetis vernus is common in England and Wales but has a highly localised distribution in Scotland and Ireland.

Cloeon dipterum (Linnaeus)
Common name: Pond Olive

Life size

Female sub-imago

Tails of similar length. Ends tend to bend downwards

More than five dark rings on each tail between abdomen and median dark band

Median black band

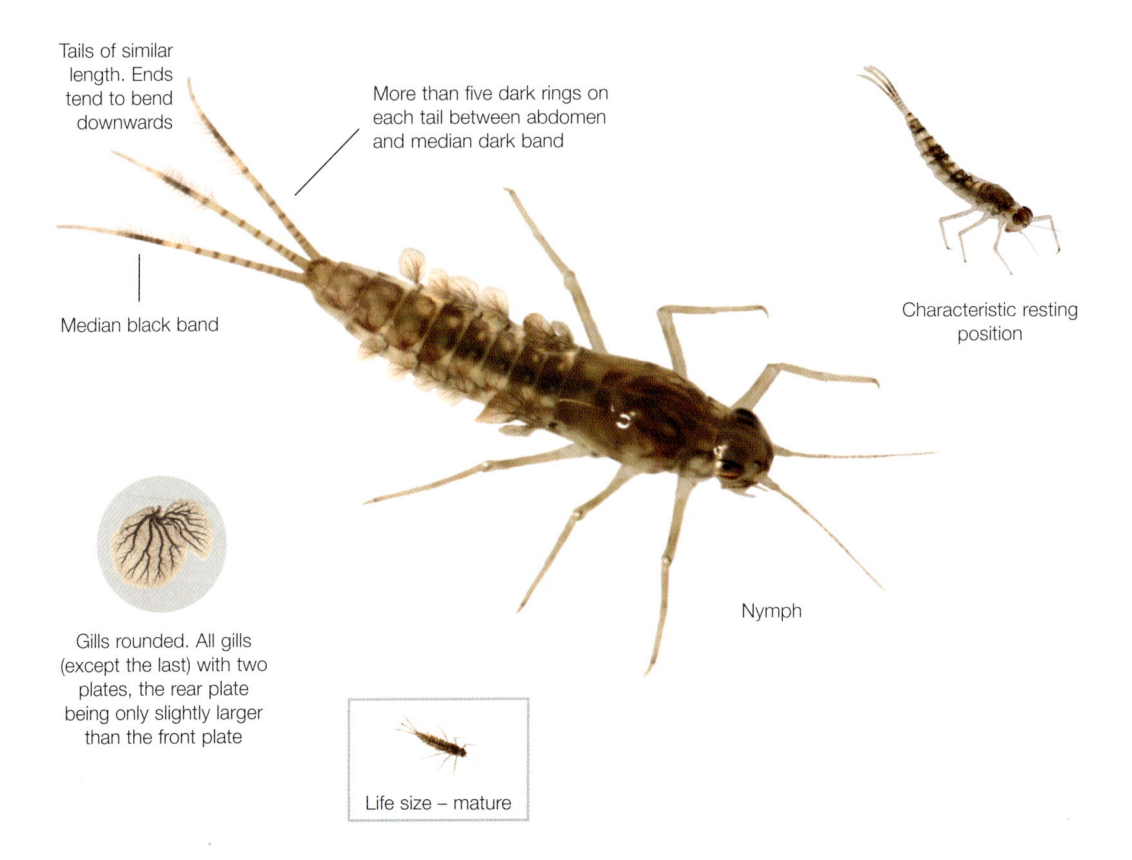

Characteristic resting position

Nymph

Gills rounded. All gills (except the last) with two plates, the rear plate being only slightly larger than the front plate

Life size – mature

Cloeon dipterum (Linnaeus)
Common name: Pond Olive

Key features

Nymphs: Streamlined nymphs with 7 pairs of plate-like gills, six of which are double gills. Tails all the same length and with a median black band.

Adults: Medium sized flies with two tails and no hindwings. The male sub-imago has dull orange-brown eyes with two faint red lines across their centre. The body is grey-olive, the last three segments of which are grey-brown. The wings are pale grey and the legs are pale white with faint reddish marks on the top of forelegs. The female has dull green eyes with two faint red lines across their centre. The body is dark brownish-olive with reddish streaks. The legs are a watery olive colour and the top of the forelegs are often markedly reddish. In both the male and female the tails are pale grey and heavily ringed at each segment with dark brown.

In the female imago the wings are transparent with reddish-brown veins. The females of this species have a distinct yellow leading edge to the wing. Their bodies vary between an apricot colour, streaked with red to a reddish colour tinged with yellow and with two faint parallel lines on the underside. The legs are bright olive-green and the forelegs often have a ring of red around them. The eyes of the female are black, while in the male they are orange-red with two faint parallel red lines through their centre. The male imago has transparent wings with distinctive brown veins along the leading edges. Its body is dull translucent cream with the last three segments much darker. The legs are pale grey-white and the tails are ringed with brown.

Separating from other species

Nymphs: *Cloeon dipterum* is similar to *C. simile*, however the two species can be separated by the shape and size of the gills. In *C. dipterum* the plates of the double gills are of a similar size, whereas in *C. simile* one plate is much bigger than the other. In addition, in *C. dipterum* the tips of the gills are rounded while in *C. simile* the larger gill has pointed tips. Where the gills have been lost, the mouth parts can be used to separate these species. You should consult a specialist key for further information.

Adults: *Cloeon* species can be separated from other Baetidae by the absence of hindwings. The only other Baetidae without hindwings is *Procloeon bifidum*. In *Cloeon* species the first segment of the foot is about twice the length of the second, whereas in *P. bifidum* the first segment is about three times the length of the second. In addition, the major cross-veins in the upper portion of the forewing are offset in *Cloeon* species while in *P. bifidum* these cross-veins line up.

Cloeon dipterum can be separated from *C. simile* by the number of small cross-veins at the tip of the forewing. In *C. dipterum* there are only three to five cross-veins, whereas in *C. simile* there are nine to eleven cross-veins. In addition, female specimens of *C. dipterum* have a distinctive brownish-yellow leading edge to the forewing. In *C. simile* the leading edge is uncoloured.

Habitat and Ecology

Nymphs of this species live in pools and margins of rivers and streams or in eutrophic ponds or shallow water in larger lakes. The nymphs swim in short, darting bursts amongst the substrate, or climb amongst the vegetation. They feed by scraping algae from submerged stones and other structures, or by gathering or collecting fine particulate organic detritus from the sediment.

There are two or more generations a year, one of which overwinters as nymphs and emerges in the spring. Adults are generally seen between May and October, but in warm weather they may also be found in other months.

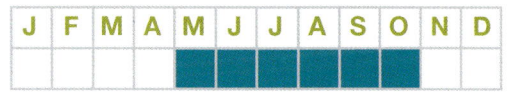

J	F	M	A	M	J	J	A	S	O	N	D
				■	■	■	■	■	■		

Emergence of the adults typically takes place at the surface during daylight hours and at dusk, with the males swarming in the afternoon or at dusk.

Unusually for Ephemeroptera, the female of this species rests for up to 14 days after mating. She then lays around 1,200 eggs directly on the water surface where they hatch and the nymphs swim away.

Distribution

Cloeon dipterum is a common and widespread species that is found throughout the British Isles. It is thought that this species is the more common of the British *Cloeon* species.

Cloeon simile Eaton
Common name: Lake Olive

Life size

Female sub-imago

Cloeon simile Eaton
Common name: Lake Olive

Key features

Nymphs: Streamlined nymphs with 7 pairs of plate-like gills, some of which are double plates. Tails all the same length and with a median black band.

Adults: Medium sized flies with two tails and no hindwings. The sub-imago has smoky grey wings with yellowish veins along the leading edges. The body is pale reddish or grey-brown, with the hind edges of each segment being slightly paler. The tails are dark grey and the legs are typically olive-green with the feet a much darker shade, sometimes appearing black. In males the eyes are olive-brown, while in females they are olive with brown spots.

The imago has transparent wings with faint touches of yellow along the leading edges. The female has a deep chestnut-brown coloured body while in the male it is pale brown-olive, with the last three segments a deeper red-brown colour. In both sexes the legs are yellow and the feet are dark, almost black. The tails are white with each segment ringed faintly with red and the eyes are yellow or olive-green.

Separating from other species

Nymphs: *Cloeon simile* is similar to *C. dipterum*, however the two species can be separated by the shape and size of the gills. In *C. dipterum* the plates of the double gills are of a similar size, whereas in *C. simile* one plate is much bigger than the other. In addition, in *C. dipterum* the tips of the gills are rounded, while in *C. simile* the larger gill has pointed tips.

Adults: *Cloeon* species can be separated from other Baetidae by the absence of hindwings. The only other Baetidae without hindwings is *Procloeon bifidum*. In *Cloeon* species the first segment of the foot is about twice the length of the second, whereas in *P. bifidum* the first segment is about three times the length of the second. In addition, the major cross-veins in the upper portion of the forewing are offset in *Cloeon* species while in *P. bifidum* these cross-veins line up.

Cloeon simile can be separated from *C. dipterum* by the number of small cross-veins at the tip of the forewing. In *C. simile* there are nine to eleven cross-veins, whereas in *C. dipterum* there are only three to five cross-veins. In addition, female specimens of *C. dipterum* have a distinctive brownish-yellow leading edge to the forewing. In *C. simile* the leading edge is uncoloured.

Habitat and Ecology

Nymphs of this species live in pools and margins of rivers and streams or in deeper water in larger lakes. The nymphs swim in short, darting bursts amongst the substrate or climb amongst the vegetation. They feed by scraping algae from submerged stones and other structures, or by gathering or collecting fine particulate organic detritus from the sediment.

There are usually two generations a year, one of which overwinters as nymphs and emerges in the spring. This species has a particularly long flight period with adults generally seen between March and November.

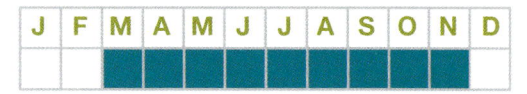

J	F	M	A	M	J	J	A	S	O	N	D
		■	■	■	■	■	■	■	■	■	

Emergence of the adults typically takes place at the surface during daylight hours, with the males of this species swarming during the morning and afternoon. Once mated, the female descends to the surface of the water and releases a few eggs by dipping the tip of her abdomen onto the water surface at intervals, or by actually settling on the surface for short periods. After several visits to the water her supply of up to 3,500 eggs is finished and the spent female falls on to the surface.

Distribution

Cloeon simile is a common and widespread species that is found throughout the British Isles. It is thought that this species is the least common of the British *Cloeon* species.

Centroptilum luteolum (Müller)
Common name: Small Spurwing

Life size

Male

Female sub-imago

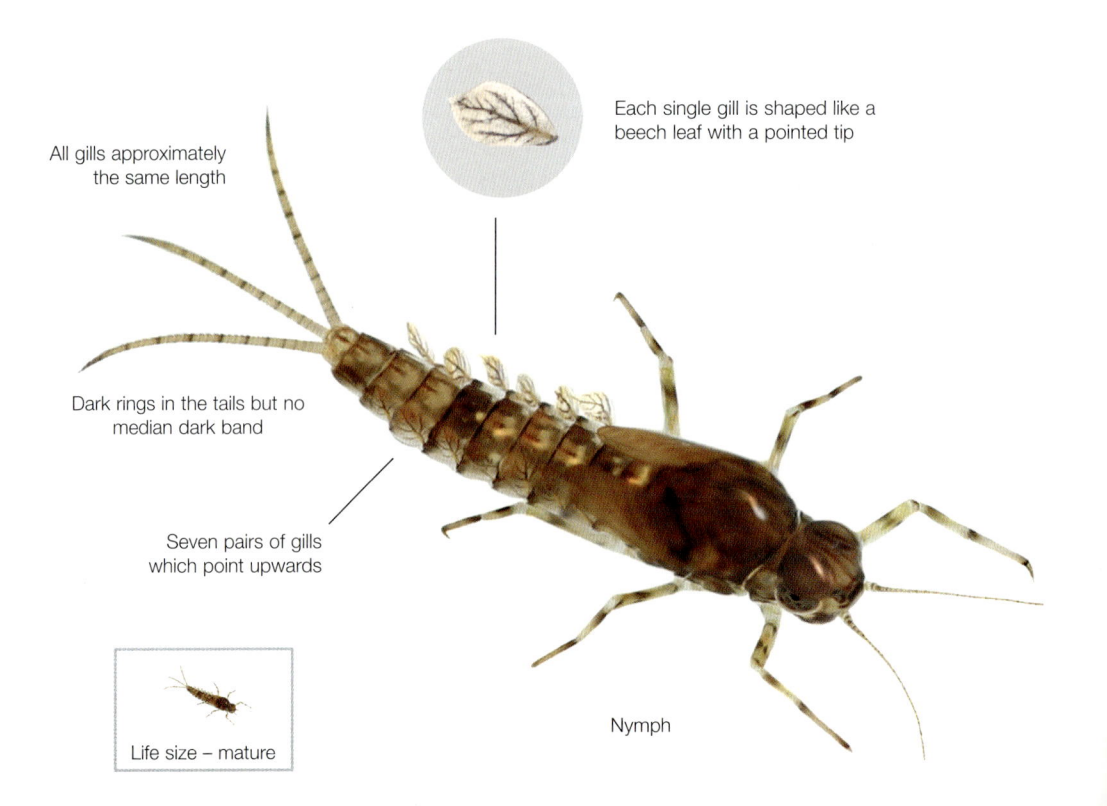

All gills approximately the same length

Each single gill is shaped like a beech leaf with a pointed tip

Dark rings in the tails but no median dark band

Seven pairs of gills which point upwards

Life size – mature

Nymph

Centroptilum luteolum (Müller)
Common name: Small Spurwing

Key features

Nymphs: Streamlined nymphs with 7 pairs of plate-like gills which are kept upright above the body. Tails all the same length and with dark rings but no median black band.

Adults: Small flies with two tails and small spur-shaped hindwings. The male sub-imago has bright orange-red eyes and a pale olive-grey body. The legs are brown-olive and the feet are greyish. In contrast, the female has pale green eyes and a pale watery brownish-olive body. The legs are a paler shade of olive than the male, however the feet are greyish as in the male. In both sexes the wings vary from pale to blue-grey, and the tails are also grey.

The male imago has bright orange-red eyes and a translucent, watery white body, the last three segments of which are pale orange-brown. In the female the eyes are dark brown and the body is two-tone: the top half is pale amber, while the underside is pale creamy-yellow. In both sexes the legs are pale olive.

Separating from other species

Nymphs: *Centroptilum luteolum* can be separated from *Baetis* species by the length of the tails. In *C. luteolum* all three tails are approximately the same length whereas in the other species the middle tail is shorter than the outer tails.

To separate *C. luteolum* from the remaining British Baetidae species you must examine the gills. *C. luteolum* has seven pairs of gills, each of which consists of a single plate shaped like a beech leaf and with a pointed tip. *Procloeon bifidum* also has seven pairs of single gills but the gills never have pointed tips.

Adults: *Centroptilum luteolum* and *Procloeon pennulatum* can be separated from other British mayflies by their spur-shaped hindwings. The hindwings are very small and are easily overlooked. In *C. luteolum* the spur-shaped hind wing has a pointed tip while in *P. pennulatum* the tip is rounded.

Habitat and Ecology

Nymphs of this species live in pools and margins of rivers and streams or on the wave lashed shores of larger stillwaters. The nymphs swim in short, darting bursts amongst the substrate or climb amongst the vegetation. They feed by scraping algae from submerged stones and other structures, or by gathering or collecting fine particulate organic detritus from the sediment.

There are usually two generations a year, one of which overwinters as nymphs and emerges in the spring, and a fast growing summer generation that emerges during the summer. This results in a fairly long flight period with adults present between April and November.

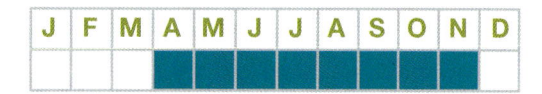

J	F	M	A	M	J	J	A	S	O	N	D
		■	■	■	■	■	■	■	■	■	

Emergence of the adults typically takes place at the surface during daylight hours, with the males swarming just above the surface of the water throughout the day and also at dusk. Once mated, the female fly descends to the surface of the water where she releases around 2,500 eggs in a single batch by dipping the tip of her abdomen onto the water surface. After releasing her eggs the spent female falls on to the surface.

Distribution

Centroptilum luteolum is a widespread and common species, found throughout the British Isles. It has a more localised distribution in Scotland, where it is more often encountered in standing waters rather than running waters.

Procloeon bifidum (Bengtsson)
Common name: Pale Evening Dun

Procloeon bifidum (Bengtsson)
Common name: Pale Evening Dun

Key features

Nymphs: Streamlined nymphs with 7 pairs of plate-like gills, all of which are single. Tails all the same length and with a narrow black band.

Adults: Small flies with two tails and no hindwings. The sub-imago has pale grey wings, a pale straw coloured body and olive-grey tails. The legs are pale olive, becoming lighter closer to the feet. The eyes are dark olive-green in the female and yellow in the male.

The imago is very similar to that of *Baetis fuscatus*, however it has a distinctive greenish tinge to the leading edge of the forewings. In addition, the hindwing is absent in *Procloeon bifidum* and there are six to eight small cross-veins along the top of the forewing.

Separating from other species

Nymphs: *Procloeon bifidum* can be separated from *Baetis* species by the length of the tails. In *P. bifidum* all three tails are approximately the same length whereas in the other species the middle tail is shorter than the outer tails.

To separate *P. bifidum* from the remaining British Baetidae species you must examine the gills. *P. bifidum* has seven pairs of gills, each of which consists of a single plate which is markedly asymmetrical with a rounded tip. *Centroptilum luteolum* also has seven pairs of single gills, however the gills are more symmetrical and shaped like a beech leaf with a pointed tip.

Adults: *Procloeon bifidum* can be separated from other Baetidae by the absence of hindwings. The only other Baetidae without hindwings are *Cloeon dipterum* and *C. simile*. In *Cloeon* species the first segment of the foot is about twice the length of the second, whereas in *P. bifidum* the first segment is about three times the length of the second. In addition, the major cross-veins in the upper portion of the forewing are offset in *Cloeon* species while in *P. bifidum* these cross-veins line up.

Habitat and Ecology

Nymphs of this species live in pools and margins of rivers and streams where they swim in short, darting bursts amongst the substrate or climb amongst the vegetation. They feed by scraping algae from submerged stones and other structures, or by gathering or collecting fine particulate organic detritus from the sediment.

Little is known about the life cycle of this species however in Central Europe there is more than one generation a year, one of which overwinters as eggs. Adults of this species can be seen between April and October.

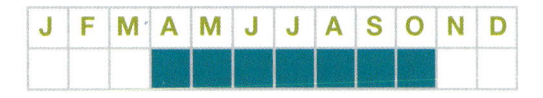

J	F	M	A	M	J	J	A	S	O	N	D
			■	■	■	■	■	■	■		

Emergence of the adults probably takes place at the surface of the water during daylight hours, with the males swarming at dawn and dusk.

Distribution

Procloeon bifidum is found throughout the British Isles, although it appears to have a very localised distribution.

Procloeon pennulatum (Eaton)
Common name: Large Spurwing

Life size

Female sub-imago

Procloeon pennulatum (Eaton)
Common name: Large Spurwing

Key features

Nymphs: Streamlined nymphs with 7 pairs of plate-like gills, most of which are double. Tails all the same length and with a median black band.

Adults: Medium sized flies with two tails and small spur-shaped hindwings. The sub-imago of this species has dark grey-olive wings, grey tails and olive legs with greyish feet. The male has dull orange eyes and a pale olive-brown body. In some specimens the last three segments appear to be more amber coloured. The female has yellow-green eyes and a pale olive to creamy-grey body. In both sexes the wings of the sub-imago are splayed on emergence, so that the angle between them is roughly 90 degrees. This is characteristic of this species.

The imago has transparent wings, two pale grey tails and pale olive-grey legs. The male has bright orange eyes, while in the female they are darker yellow-green. In the male the body is translucent white with each segment ringed with pale red. The last three segments are dark amber. In contrast, the body of the female is pale olive and is covered with dark reddish-amber flecks.

Separating from other species

Nymphs: Six of the seven pairs of gills in *Procloeon pennulatum* consist of double plates. The seventh pair has single plates. Both *Cloeon simile* and *P. pennulatum* have one of the double plates much large than the other, however in *C. simile* the large plate has a pointed tip whereas in *P. pennulatum* the larger plate is rounded. A further characteristic to separate *P. pennulatum* from the British *Cloeon* species is the number of dark rings on the tails. In *P. pennulatum* there are no more than five dark rings between the body and the dark band on the tails, while in *Cloeon* species there are more than 5 dark rings. Finally, unlike other Baetidae, *P. pennulatum* holds its tails very closely together when at rest.

Adults: *Procloeon pennulatum* and *Centroptilum luteolum* can be separated from other British mayflies by their spur-shaped hindwings. The hindwings are very small and are easily overlooked. In *P. pennulatum* the spur-shaped hindwing has a rounded tip while in *C. luteolum* the tip is pointed.

Habitat and Ecology

Nymphs of this species live in pools and margins of rivers and streams where they swim in short, darting bursts amongst the substrate or climb amongst the vegetation. They feed by scraping algae from submerged stones and other structures, or by gathering or collecting fine particulate organic detritus from the sediment.

There is one generation a year, which overwinters as eggs, however some workers have suggested that there may be a second generation in the summer. Adults of this species can be seen between May and October.

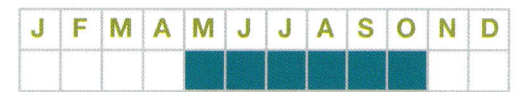

J	F	M	A	M	J	J	A	S	O	N	D
				■	■	■	■	■	■		

Emergence of the adults probably takes place at the surface of the water during daylight hours, with the males swarming at dusk. Once mated, the female probably lays her 2,500 eggs in either a single batch or several batches by dipping the tip of her abdomen onto the water surface.

Distribution

Procloeon pennulatum is known from England and Wales, but is rarely encountered in Scotland and absent from Ireland. As a result, a voucher specimen is required for records from these areas.

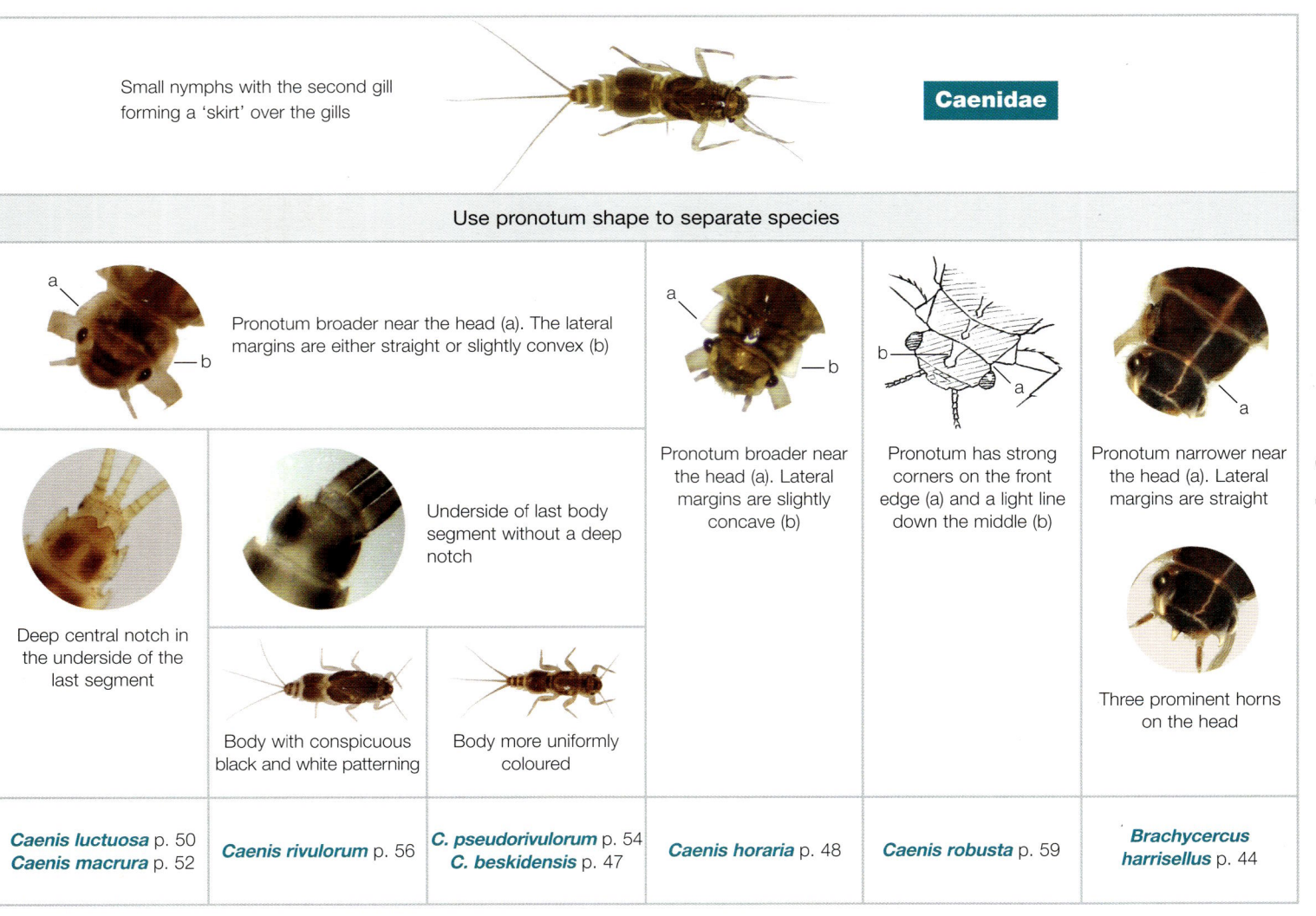

Identification Chart 3A: Caenidae – Nymphs

Caenidae

Small nymphs with the second gill forming a 'skirt' over the gills

Use pronotum shape to separate species

	Pronotum narrower near the head (a). Lateral margins are straight. Three prominent horns on the head	*Brachycercus harrisellus* p. 44
	Pronotum has strong corners on the front edge (a) and a light line down the middle (b)	*Caenis robusta* p. 59
	Pronotum broader near the head (a). Lateral margins are slightly concave (b)	*Caenis horaria* p. 48
	Pronotum broader near the head (a). The lateral margins are either straight or slightly convex (b). Underside of last body segment without a deep notch. Body more uniformly coloured	*C. pseudorivulorum* p. 54 / *C. beskidensis* p. 47
	Body with conspicuous black and white patterning	*Caenis rivulorum* p. 56
	Deep central notch in the underside of the last segment	*Caenis luctuosa* p. 50 / *Caenis macrura* p. 52

Identification Chart 3B: **Caenidae – Adults**

Caenidae

Very small flies with three tails and no hindwings.
Forewing size: 3.0-9.0mm

Use distance between the legs to separate genera

Base of front legs close together

Caenis spp.

	Dorsal spine present *		Dorsal spine absent *	
Markings on segments 1-3 only	Markings on segments 1-5 or 6	Markings on all segments	Spines at edge of body segment 9 are longer than they are broad	Spines at edge of body segment 9 are not longer than they are broad
Caenis rivulorum p. 56	*Caenis horaria* p. 48	*Caenis robusta* p. 59	*Caenis luctuosa* p. 50 *Caenis macrura* p. 52 *Caenis pusilla* p. 55	*C. beskidensis* p. 47 *C. pseudorivulorum* p. 54

Base of the front legs (look underneath) of your specimen widely separated so that the legs attach to the body at the outer margin of the underside

Brachycercus harrisellus p. 44

* The dorsal spine is very small. It is located on the 2nd abdominal segment. The spine is best viewed from the side

Brachycercus harrisellus Curtis
Common name: Large Broadwings

Lateral margins of abdominal segments 3-7 have flat backwards-facing projections

Head has three prominent horn-like protuberances (ocellar tubercles)

Nymph

Life size – mature

Brachycercus harrisellus Curtis
Common name: Large Broadwings

Key features

Nymphs: Small nymphs (6-11mm). Gills 3 to 7 covered with two large plates. Three prominent horn-like protuberances on the head.

Adults: Small flies with three tails and no hindwings. The adults of this species have a dark, almost black body and broad milky-white wings. The tails are much longer than the body, which is relatively short.

Separating from other species

Nymphs: *Brachycercus harrisellus* is easily separated from other British Caenidae by the presence of three horn-like protuberances on the head.

Adults: *Brachycercus harrisellus* can be separated from other British Caenidae by the distance between the front pair of legs. In *Brachycercus* the front legs join the body at the edges, whereas in *Caenis* the front legs are attached closer to the centre of the body.

Habitat and Ecology

Nymphs of this species live in the pools and margins of rivers and streams, where they burrow into, and live on mud and silt on the bed of the watercourse. They are poor swimmers but are adapted for moving amongst mud and silt, where they feed by collecting or gathering fine particulate organic detritus from the sediment.

There is one generation a year, which usually overwinters as eggs and emerges in July.

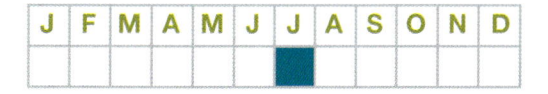

Distribution

This species has a widespread, though patchy distribution throughout the country. The most northerly records are from the River Forth in Stirlingshire and the River Black Devon in Clackmannanshire.

J	F	M	A	M	J	J	A	S	O	N	D
						■					

Caenis beskidensis Sowa
Common name: Anglers' Curse

Key features

Nymphs: Very small nymphs (4-9mm). Gills 3 to 7 covered with two large plates.

Adults: Very small flies with three tails and no hindwings. *Caenis* have very broad wings, which are fringed with hairs in both the sub-imago and the imago. The wing venation is very simple with very few cross-veins. The body varies between black and light brown and contrasts markedly with the wings, which are typically pale or white in colour. The tails of male imago are many times longer than the body.

Separating from other species

Nymphs: *Caenis beskidensis* is superficially similar to both *C. pseudorivulorum* and *C. rivulorum*. Separation of these species relies on minute differences in the shape of the body segments and the head. It is recommended that a specialist key be used to separate these species.

Adults: *C. beskidensis* is superficially similar to *C. pseudorivulorum* and identification relies upon minute differences in the male genitalia. It is recommended that a specialist key be consulted to separate this species from other *Caenis* species.

Habitat and Ecology

Little is known about this species, however it is likely that the nymphs live in the pools and margins of stony rivers and streams, in silt trapped between larger stones and gravel. They are poor swimmers but are adapted for moving amongst the sediment. They feed by collecting or gathering fine particulate organic detritus from the sediment.

There is one generation a year, which usually overwinters as nymphs and emerges between July and September.

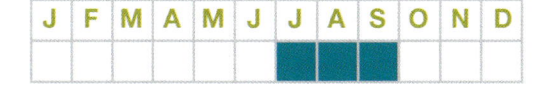

Distribution

Caenis beskidensis was first recorded from the River Crûg, Powys in 1984. Since then there have been no other records, however this may be as a result of under-recording due to the superficial resemblance of its nymphs to the common British species *Caenis rivulorum*. A voucher specimen is required for all records of this species.

J	F	M	A	M	J	J	A	S	O	N	D
						■	■	■			

Caenis horaria (Linnaeus)
Common name: Anglers' Curse

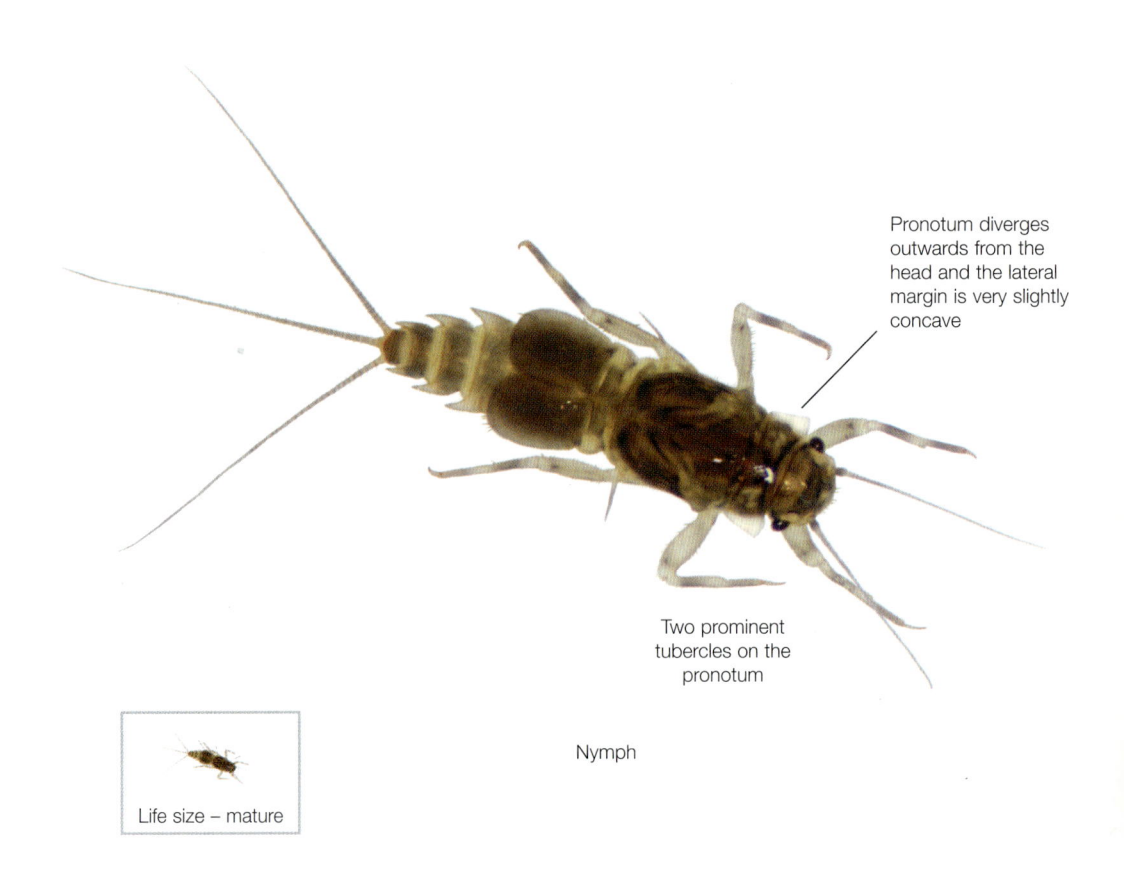

Pronotum diverges outwards from the head and the lateral margin is very slightly concave

Two prominent tubercles on the pronotum

Nymph

Life size – mature

Caenis horaria (Linnaeus)
Common name: Anglers' Curse

Key features

Nymphs: Very small nymphs (4-9mm). Gills 3 to 7 covered with two large plates.

Adults: Very small flies with three tails and no hindwings. *Caenis* have very broad wings, which are fringed with hairs in both the sub-imago and the imago. The wing venation is very simple with very few cross-veins. The body varies between black and light brown and contrasts markedly with the wings, which are typically pale or white in colour. In addition, there are grey markings on segments 1 to 5 or sometimes segments 1 to 6 of the body. The tails of male imago are many times longer than the body.

Separating from other species

Nymphs: *Caenis horaria* can be separated from other *Caenis* species by the shape of the pronotum, which is broader near the head with slightly concave edges, unlike those of other species, which are either straight or slightly convex. In addition there are two prominent tubercles or 'bumps' on the pronotum.

Adults: Adults of the most common British Caenidae fall into two broad groups. The first group, which contains *Caenis horaria*, *C. rivulorum* and *C. robusta*, consists of flies with plain white or yellowish-white tails. In addition, there is a small spine projecting backwards from the upper surface of the second body segment. The second group, which contains *C. luctuosa*, *C. macrura*, *C. beskidensis*, *C. pseudorivulorum* and *C. pusilla*, does not have a spine on the upper surface of the second body segment and has brownish-grey or greyish-white tails which sometimes have darker rings on them.

Caenis horaria can be separated from *C. rivulorum* and *C. robusta* by the pattern of markings on the upper surface of the body. In *C. horaria* there are greyish markings on body segments 1 to 5 or sometimes 1 to 6. In *C. rivulorum* these markings are on segments 1 to 3 only while in *C. robusta* the markings are present on all the body segments.

Habitat and Ecology

Nymphs of this species live in the pools and margins of rivers and streams or in lakes and canals, where they burrow into and live on mud and silt on the bed of the watercourse. The nymphs are poor swimmers but are adapted for moving amongst mud and silt where they feed by collecting or gathering fine particulate organic detritus from the sediment.

There is typically one generation a year, which overwinters as nymphs, although some workers have suggested that there may be a further summer generation. Adults have been recorded between May and September.

J	F	M	A	M	J	J	A	S	O	N	D
				■	■	■	■	■			

Emergence of the adults typically takes place partially or entirely out of the water on a stick, stone or plant stem at dusk. The males of this species can be found swarming at dusk and probably during the night.

Distribution

This species has a widespread, though localised, distribution. The majority of records for this species are from the south-east of England, however there are records from as far north as Sutherland.

Caenis luctuosa (Bürmeister)
Common name: Anglers' Curse

Pronotum broader near the head. The lateral margin is either straight or slightly convex

Last visible sternite has a deep central notch in its posterior margin

Nymph

Life size – mature

Caenis luctuosa (Bürmeister)
Common name: Anglers' Curse

Key features

Nymphs: Very small nymphs (4-9mm). Gills 3 to 7 covered with two large plates.

Adults: Very small flies with three tails and no hindwings. *Caenis* have very broad wings, which are fringed with hairs in both the sub-imago and the imago. The wing venation is very simple with very few cross-veins. The body varies between black and light brown and contrasts markedly with the wings, which are typically pale or white in colour. The tails of male imago are many times longer than the body.

Separating from other species

Nymphs: *Caenis luctuosa* and *C. macrura* can be separated from other *Caenis* species by the presence of a deep notch in the underside of the last body segment and the shape of the pronotum, which is broader near the head and has either straight or slightly convex edges. Separation of nymphs of *C. luctuosa* and *C. macrura* is particularly difficult and unreliable, although some workers believe that differences in the colouration of the body segments can be used to separate these species.

Adults: Adults of the most common British Caenidae fall into two broad groups. The first group, which contains *Caenis horaria*, *C. rivulorum* and *C. robusta*, consists of flies with plain white or yellowish-white tails. In addition, there is a small spine projecting backwards from the upper surface of the second body segment. The second group, which contains *C. luctuosa*, *C. macrura*, *C. beskidensis*, *C. pseudorivulorum* and *C. pusilla*, does not have a spine on the upper surface of the second body segment and has brownish-grey or greyish-white tails which sometimes have darker rings on them.

Caenis luctuosa and *C. macrura* can be separated from *C. beskidensis*, *C. pseudorivulorum* and *C. pusilla* by the length of the spines on the side of the ninth body segment. In *C. luctuosa* and *C. macrura* these spines are longer than they are broad. In *C. beskidensis*, *C. pseudorivulorum* and *C. pusilla* they are never longer than they are broad.

Caenis luctuosa can be separated from *C. macrura* by the shape of the antenna. In *C. luctuosa* the filament is fat at the base so that the filament narrows abruptly close to the base before tapering to a point. In *C. macrura* the antenna filament is thin at the base so that the filament tapers slowly to a point.

Habitat and Ecology

Nymphs of this species live in the pools and margins of rivers or on lake shores, where they live in silt trapped between larger stones and gravel. The nymphs are poor swimmers but are adapted for moving amongst the sediment. They feed by collecting or gathering fine particulate organic detritus from the sediment.

There is either one or two generations a year, one of which overwinters as nymphs. Adults have been recorded between June and September.

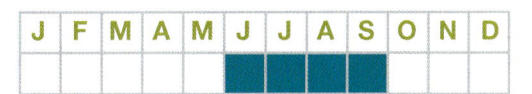

J	F	M	A	M	J	J	A	S	O	N	D
					■	■	■	■			

Emergence of the adults typically takes place partially or entirely out of the water on a stick, stone or plant stem at dusk. The males of this species can be found swarming at dawn.

Distribution

This species has a widespread distribution, although it is scarce in the north of Scotland. The majority of records for this species are from the south-east of England, however due to the problems with the identification of nymphs, it is difficult to compile a complete distribution for this species. *Caenis luctuosa* is thought to be more common than *C. macrura*.

Caenis macrura Stephens
Common name: Anglers' Curse

Life size

No dilation at base of terminal bristle of antenna – separates from *C. luctuosa*

Sub-imago

Transposing imago

Caenis macrura Stephens
Common name: Anglers' Curse

Key features

Nymphs: Very small nymphs (4-9mm). Gills 3 to 7 covered with two large plates.

Adults: Very small flies with three tails and no hindwings. *Caenis* have very broad wings, which are fringed with hairs in both the sub-imago and the imago. The wing venation is very simple with very few cross-veins. The body varies between black and light brown and contrasts markedly with the wings, which are typically pale or white in colour. The tails of male imago are many times longer than the body.

Separating from other species

Nymphs: *Caenis macrura* and *C. luctuosa* can be separated from other *Caenis* species by the presence of a deep notch in the underside of the last body segment and the shape of the pronotum, which is broader near the head and has either straight or slightly convex edges. Separation of nymphs of *C. macrura* and *C. luctuosa* is particularly difficult and unreliable, although some workers believe that differences in the colouration of the body segments can be used to separate these species.

Adults: Adults of the most common British Caenidae fall into two broad groups. The first group, which contains *Caenis horaria*, *C. rivulorum* and *C. robusta*, consists of flies with plain white or yellowish-white tails. In addition, there is a small spine projecting backwards from the upper surface of the second body segment. The second group, which contains *C. luctuosa*, *C. macrura*, *C. beskidensis*, *C. pseudorivulorum* and *C. pusilla*, does not have a spine on the upper surface of the second body segment and has brownish-grey or greyish-white tails which sometimes have darker rings on them.

Caenis macrura and *C. luctuosa* can be separated from *C. beskidensis*, *C. pseudorivulorum* and *C. pusilla* by the length of the spines on the side of the ninth body segment. In *C. macrura* and *C. luctuosa* these spines are longer than they are broad. In *C. beskidensis*, *C. pseudorivulorum* and *C. pusilla* they are never longer than they are broad.

Caenis macrura can be separated from *C. luctuosa* by the shape of the antenna. In *C. macrura* the antenna filament is thin at the base so that the filament tapers slowly to a point. In *C. luctuosa* the filament is fat at the base so that the filament narrows abruptly close to the base before tapering to a point.

Habitat and Ecology

Nymphs of this species live in the pools and margins of rivers where they live in silt trapped between larger stones and gravel. The nymphs are poor swimmers but are adapted for moving amongst the sediment. They feed by collecting or gathering fine particulate organic detritus from the sediment.

There is one generation a year, which usually overwinters as nymphs and emerges between May and August.

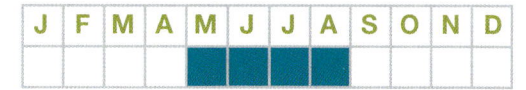

J	F	M	A	M	J	J	A	S	O	N	D
				■	■	■	■				

Emergence of the adults typically takes place partially or entirely out of the water on a stick, stone or plant stem in the early morning. The males of this species can be found swarming at dawn. Once mated, the female flies upstream and descends to the surface of the water to release a few eggs by dipping the tip of her abdomen onto the water surface at intervals, or by actually settling on the surface for short periods. After several visits to the water her supply of around 1,200 eggs is finished and the spent female falls on to the surface.

Distribution

This species has a widespread but localised distribution with the most northerly record from East Lothian. The majority of records for this species are from the south-east of England, however due to the problems with the identification of nymphs, it is difficult to compile a complete distribution for this species. *Caenis macrura* is thought to be less common than *C. luctuosa*.

Caenis pseudorivulorum Keffermüller
Common name: Anglers' Curse

Key features

Nymphs: Very small nymphs (4-9mm). Gills 3 to 7 covered with two large plates.

Adults: Very small flies with three tails and no hindwings. *Caenis* have very broad wings, which are fringed with hairs in both the sub-imago and the imago. The wing venation is very simple with very few cross-veins. The body varies between black and light brown and contrasts markedly with the wings, which are typically pale or white in colour. The tails of the male imago are many times longer than the body.

Separating from other species

Nymphs: *Caenis pseudorivulorum* is superficially similar to both *C. rivulorum* and *C. beskidensis*. Separation of these species relies on minute differences in the shape of the body segments and the head. It is recommended that a specialist key be used to separate these species.

Adults: Adults of the most common British Caenidae fall into two broad groups. The first group, which contains *Caenis horaria*, *C. rivulorum* and *C. robusta*, consists of flies with plain white or yellowish-white tails. In addition, there is a small spine projecting backwards from the upper surface of the second body segment. The second group, which contains *C. luctuosa*, *C. macrura*, *C. beskidensis*, *C. pseudorivulorum* and *C. pusilla*, does not have a spine on the upper surface of the second body segment and has brownish-grey or greyish-white tails which sometimes have darker rings on them.

   C. pseudorivulorum is superficially similar to *C. beskidensis* and identification relies upon minute differences in the male genitalia. It is recommended that a specialist key be consulted to separate this species from other *Caenis* species.

Habitat and Ecology

Little is known about this species, however it is likely that the nymphs live in the pools and margins of stony rivers and streams, in silt trapped between larger stones and gravel. They are poor swimmers but are adapted for moving amongst the sediment, where they feed by collecting or gathering fine particulate organic detritus from the sediment. There are possibly two generations per year – a slow growing winter generation and a much faster summer generation. This results in a long flight period with adults being seen from the middle of June to the end of October.

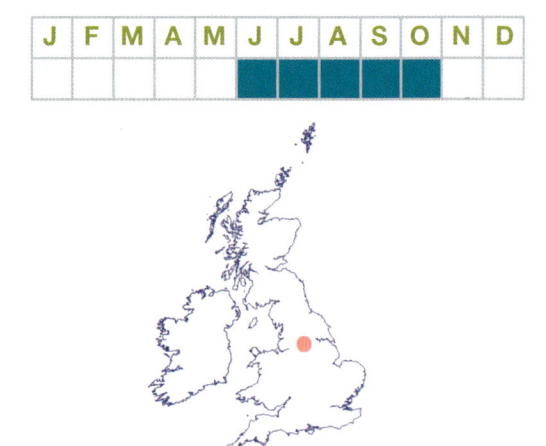

J	F	M	A	M	J	J	A	S	O	N	D
					■	■	■	■	■		

Distribution

Caenis pseudorivulorum was first recorded from the River Derwent, Yorkshire in 1984. Since then there have been few other records. However this may be as a result of under-recording due to the superficial resemblance of its nymphs to the common British species *Caenis rivulorum*. A voucher specimen is required for all records of this species.

Caenis pusilla Navás
Common name: Anglers' Curse

Key features

Nymphs: Very small nymphs (4-9mm). Gills 3 to 7 covered with two large plates.

Adults: Very small flies with three tails and no hindwings. *Caenis* have very broad wings, which are fringed with hairs in both the sub-imago and the imago. The wing venation is very simple with very few cross-veins. The body varies between black and light brown and contrasts markedly with the wings, which are typically pale or white in colour. The tails of the male imago are many times longer than the body.

Separating from other species

Nymphs: The identification of *Caenis pusilla* relies on microscopic examination of the spines on the body and bristles on the legs. It is recommended that a specialist key be consulted to separate this species from other *Caenis* species.

Adults: *Caenis pusilla* is superficially similar to a number of other *Caenis* spp. and identification relies upon minute differences in the male genitalia. It is recommended that a specialist key be consulted to separate this species from other *Caenis* species.

Habitat and Ecology

Nymphs of this species live in the pools and margins of stony rivers and streams. They are poor swimmers but are adapted for moving amongst the sediment, where they feed by collecting or gathering fine particulate organic detritus from the sediment.

There is one generation a year, which usually overwinter as nymphs and emerges between June and July.

J	F	M	A	M	J	J	A	S	O	N	D
					■	■					

Emergence of the adults typically takes place partially or entirely out of the water on a stick, stones or plant stem.

Distribution

Caenis pusilla was first recorded from the River Itchen, Hampshire in the early 1980s. Since then there have been few other records, however this may be as a result of under-recording due to the superficial resemblance of its nymphs to the common British species *Caenis rivulorum*. A voucher specimen is required for all records of this species.

Caenis rivulorum Eaton
Common name: Anglers' Curse

Life size

Female imago

Characteristic pied appearance due to the pale abdominal segments between the thorax and gill covers

Life size – mature

Nymph

Caenis rivulorum Eaton
Common name: Anglers' Curse

Key features

Nymphs: Very small nymphs (4-9mm) with a distinctive pied appearance. Gills 3 to 7 covered with two large plates.

Adults: Very small flies with three tails and no hindwings. *Caenis* have very broad wings, which are fringed with hairs in both the sub-imago and the imago. The wing venation is very simple with very few cross-veins. The body varies between black and light brown and contrasts markedly with the wings, which are typically pale or white in colour. In addition, there are grey markings on segments 1 to 3 of the body. The tails of the male imago are many times longer than the body.

Separating from other species

Nymphs: *Caenis rivulorum* is superficially similar to both *C. pseudorivulorum* and *C. beskidensis*. Separation of these species relies on minute differences in the shape of the body segments and the head. It is recommended that a specialist key be used to separate these species.

Adults: Adults of the most common British Caenidae fall into two broad groups. The first group, which contains *Caenis horaria*, *C. rivulorum* and *C. robusta*, consists of flies with plain white or yellowish-white tails. In addition, there is a small spine projecting backwards from the upper surface of the second body segment. The second group, which contains *C. luctuosa*, *C. macrura*, *C. beskidensis*, *C. pseudorivulorum* and *C. pusilla*, does not have a spine on the upper surface of the second body segment and has brownish-grey or greyish-white tails which sometimes have darker rings on them.

Caenis rivulorum can be separated from *C. horaria* and *C. robusta* by the pattern of markings on the upper surface of the body. In *C. rivulorum* there are greyish markings on segments 1 to 3 only. In *C. horaria* these markings are on body segments 1 to 5 or sometimes 1 to 6, while in *C. robusta* the markings are present on all the body segments.

Habitat and Ecology

Nymphs of this species live in the pools and margins of stony rivers and streams, where they live in silt trapped between larger stones and gravel. The nymphs are poor swimmers but are adapted for moving amongst the sediment. They feed by collecting or gathering fine particulate organic detritus from the sediment. There is one generation a year which overwinters as nymphs and emerges as adults between May and September.

J	F	M	A	M	J	J	A	S	O	N	D
				■	■	■	■	■			

Emergence of the adults probably takes place partially or entirely out of the water on a stick, stones or plant stem at dusk or dawn.

Distribution

Caenis rivulorum is the most common and widely distributed of the British Caenidae. It can be found throughout the British Isles and Ireland, although it has a relatively localised distribution in the south-east of England.

Caenis robusta Eaton
Common name: Anglers' Curse

Key features

Nymphs: Very small nymphs (4-9mm). Gills 3 to 7 covered with two large plates.

Adults: Very small flies with three tails and no hindwings. *Caenis* have very broad wings, which are fringed with hairs in both the sub-imago and the imago. The wing venation is very simple with very few cross-veins. The body varies between black and light brown and contrasts markedly with the wings, which are typically pale or white in colour. In addition, there are grey markings on all the body segments. The tails of the male imago are many times longer than the body.

Separating from other species

Nymphs: *Caenis robusta* can be separated from other *Caenis* species by the shape of the pronotum, which has strong corners on the front edge and a light line down the middle.

Adults: Adults of the most common British Caenidae fall into two broad groups. The first group, which contains *Caenis horaria*, *C. rivulorum* and *C. robusta*, consists of flies with plain white or yellowish-white tails. In addition, there is a small spine projecting backwards from the upper surface of the second body segment. The second group, which contains *C. luctuosa*, *C. macrura*, *C. beskidensis*, *C. pseudorivulorum* and *C. pusilla*, does not have a spine on the upper surface of the second body segment and has brownish-grey or greyish-white tails which sometimes have darker rings on them.

Caenis robusta can be separated from *C. horaria* and *C. rivulorum* by the pattern of markings on the upper surface of the body. In *C. robusta* there are greyish markings on all the body segments, while in *C. horaria* these markings are on body segments 1 to 5 or sometimes 1 to 6 and in *C. rivulorum* the markings are only present on segments 1 to 3.

Habitat and Ecology

Nymphs of this species live in the pools and margins of rivers, or on lake shores or in canals, where they live in silt and mud on the bed. The nymphs are poor swimmers but are adapted for moving amongst the sediment. They feed by collecting or gathering fine particulate organic detritus from the sediment. There is either one or two generations a year, one of which overwinters as nymphs. Adults have been recorded in June and July.

J	F	M	A	M	J	J	A	S	O	N	D
					■	■					

Emergence of the adults probably takes place partially or entirely out of the water on a stick, stones or plant stem at dusk or dawn.

Distribution

This species has a widespread, though localised distribution. The majority of records for this species are from the south-east of England, however there are records from as far north as Sutherland and the Western Isles.

Identification Chart 4A: **Ephemeridae – Nymphs**

Ephemeridae			
Large cylindrical nymphs with two tusk-like projections from the head and thick, feathery gills that are held over the back	Use body markings to separate species	Markings on body segments NOT distinct triangles nor six longitudinal stripes (N.B. markings are faint or absent on body segments 1 to 6)	*Ephemera danica* p. 62
		Six longitudinal stripes on each body segment	*Ephemera lineata* p. 64
		Large dark triangular markings on each body segment	*Ephemera vulgata* p. 66

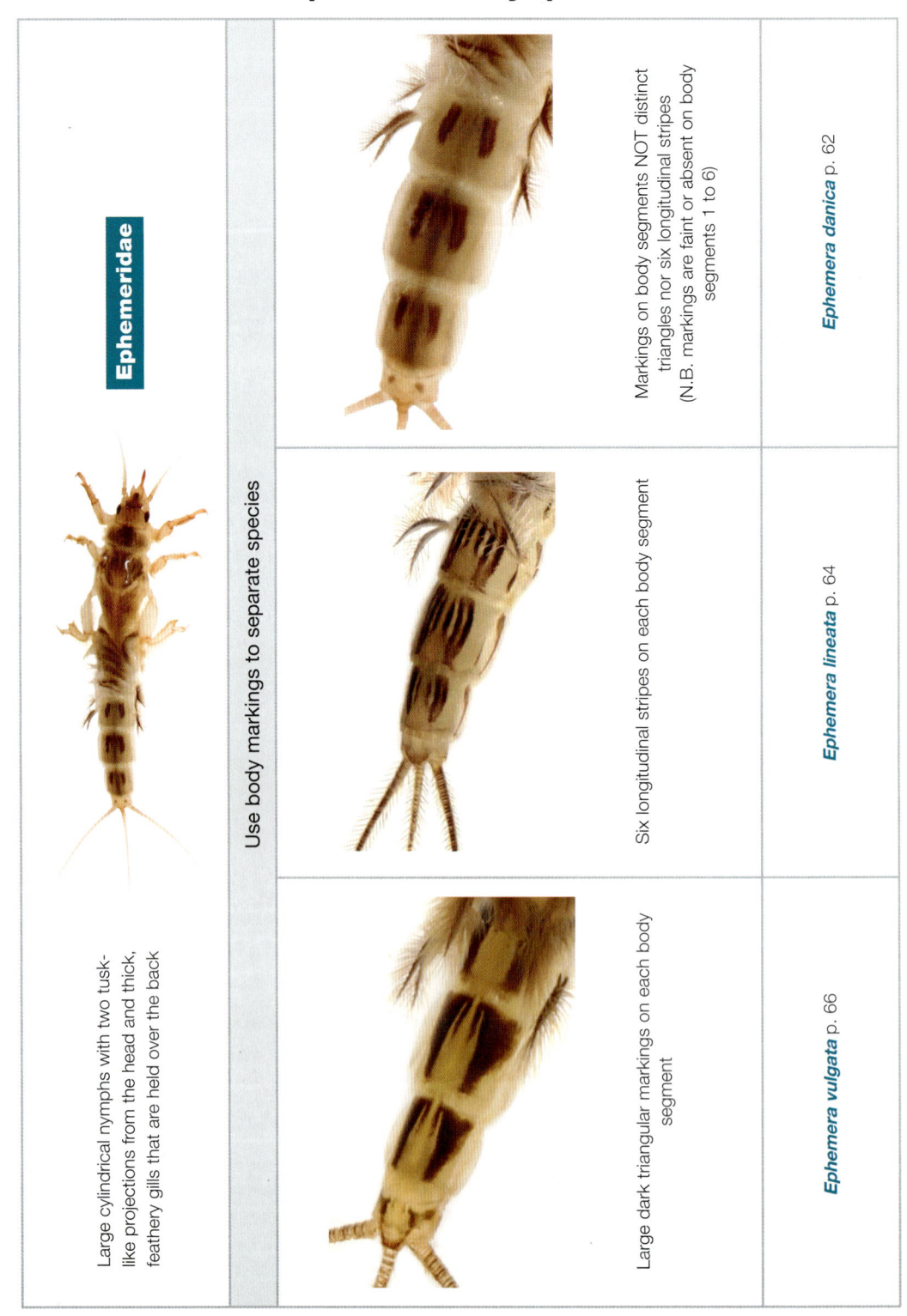

Identification Chart 4B: **Ephemeridae – Adults**

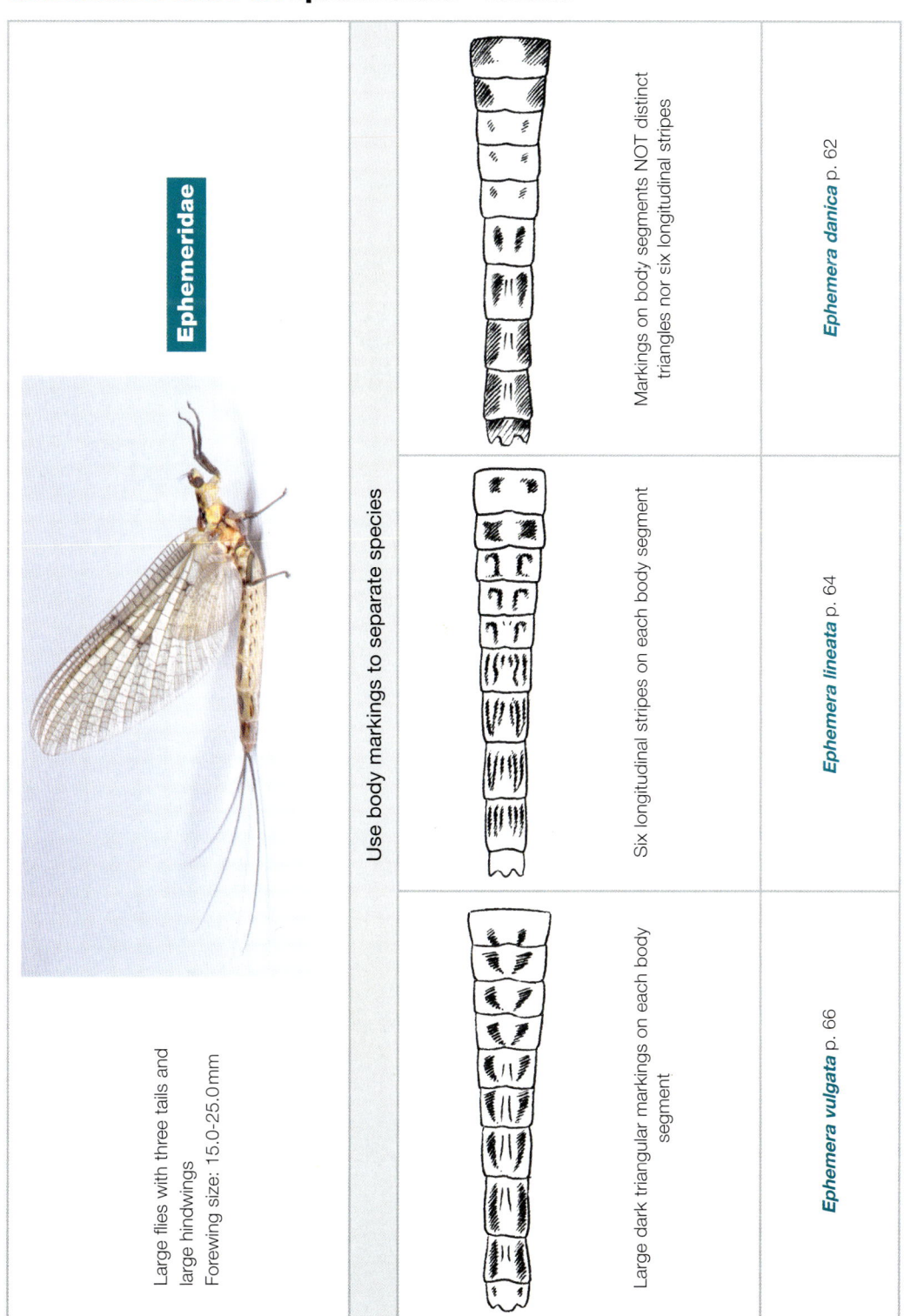

Ephemeridae

Large flies with three tails and large hindwings
Forewing size: 15.0–25.0mm

Use body markings to separate species

Markings on body segments NOT distinct triangles nor six longitudinal stripes — *Ephemera danica* p. 62

Six longitudinal stripes on each body segment — *Ephemera lineata* p. 64

Large dark triangular markings on each body segment — *Ephemera vulgata* p. 66

Ephemera danica **Müller**
Common name: Green Drake Mayfly

Life size

Female sub-imago

Markings on each segment are NOT distinct triangles (as *E. vulgata*) nor six longitudinal stripes (as *E. lineata*)

All gills thickly fringed with filaments on both sides

Nymph

Two tusk-like projections in front of head

Life size – mature

Ephemera danica Müller
Common name: Green Drake Mayfly

Key features

Nymphs: Large cylindrical nymphs with two tusk-like projections from the head and thick, feathery gills that are held over the back.

Adults: Large flies with three tails and large hindwings. The sub-imago has grey wings, with a yellowish-green tinge and heavy venation. The body is creamy-yellow with distinctive brown markings on the body segments. The tails are dark grey-black and the legs are a creamy-olive colour with black markings. In general, the colours of the wings, tails and legs are darker, and the body paler in the male.

The imago has transparent wings, which have heavy brown venation and several dark patches. The body is creamy-white with the last three segments brownish. The tails are extremely long and almost black in colour.

Separating from other species

Nymphs: The British Ephemeridae can be separated by the markings on the upper surface of their body. The markings in *Ephemera danica* are neither dark triangles nor a series of lines. On segments 7 to 9 the markings are elongated and obvious, however on the remaining segments the markings are indistinct. In *E. vulgata* there are dark triangular markings on all the body segments, although in some cases these are indistinct on the first and last segments. In *E. lineata* the markings consist of a series of six stripes on segments 7 to 9. On the remaining segments the markings are indistinct.

Adults: Adults of the British Ephemeridae can also be separated by the markings on the upper surface of their body. In *Ephemera danica* these markings are distinct on most segments but are often absent or faint on segments 1 to 5. In *E. vulgata* the markings are present on all segments and are dark and triangular or elongated, often with a pair of longitudinal stripes. *E. lineata* also has markings on each body segment. On segments 5 to 9 these consist of a series of 6 longitudinal stripes, while on other segments they are square, triangular or angular.

Habitat and Ecology

Nymphs of this species live in lakes and fast-flowing rivers and streams with a sandy or gravelly bed. They dig into the gravel to form a tubular burrow and they use their gills to force the water through this burrow. They feed by filtering or collecting fine particulate organic detritus from the water column.

Ephemera danica usually has a two year life cycle, however recent work has shown that in the warmer waters of Southern England it is able to complete its life cycle in one year. The main flight period is towards the end of May, but adults are often present between April and November.

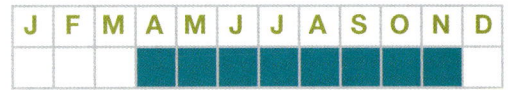

J	F	M	A	M	J	J	A	S	O	N	D
		■	■	■	■	■	■	■	■	■	

Emergence of the adults takes place during daylight on the surface of the water or occasionally on a stick, stone or plant stem partially or entirely out of the water. The males of this species can be found swarming throughout the day, and often swarming continues until dusk.

Once mated, the female flies upstream and descends to the surface of the water to release a few eggs by dipping the tip of her abdomen onto the surface at intervals, or by actually settling on the surface for short periods. After several visits to the water her supply of up to 8,300 eggs is finished and the spent female falls on to the surface.

Distribution

Ephemera danica is the most common of the British Ephemeridae. It can be found in unpolluted rivers and lakes throughout the British Isles, but its distribution is much more localised in the north.

Ephemera lineata Eaton
Common name: Striped Mayfly

Life size

Female sub-imago

Six longitudinal stripes on the abdominal tergites. This is the main identifying feature of this species

All gills thickly fringed with filaments on both sides

Nymph

Two tusk-like projections in front of head for burrowing

Life size – mature

Ephemera lineata Eaton
Common name: Striped Mayfly

Key features

Nymphs: Large cylindrical nymphs with two tusk-like projections from the head and thick, feathery gills that are held over the back.

Adults: Large flies with three tails and large hindwings. The sub-imago has grey wings, with a yellowish-green tinge and heavy venation. The body is creamy-yellow with distinctive brown markings on the body segments. The tails are dark grey-black and the legs are a creamy-olive colour with black markings. In general, the colours of the wings, tails and legs are darker, and the body paler in the male.

The imago has transparent wings, which have heavy brown venation and several dark patches. The body is creamy-white with the last three segments brownish. The tails are extremely long and almost black in colour.

Separating from other species

Nymphs: The British Ephemeridae can be separated by the markings on the upper surface of their body. The markings in *Ephemera lineata* consist of a series of six stripes on segments 7 to 9. On the remaining segments the markings are indistinct. In *E. danica* the markings are neither dark triangles nor a series of lines. On segments 7 to 9 the markings are elongated and obvious, however on the remaining segments the markings are indistinct. In *E. vulgata* there are dark triangular markings on all the body segments, although in some cases these are indistinct on the first and last segments.

Adults: Adults of the British Ephemeridae can also be separated by the markings on the upper surface of their body. These markings in *Ephemera lineata* can be found on all the segments. On segments 5 to 9 they consist of a series of 6 longitudinal stripes, while on other segments they are square, triangular or angular. In *E. danica* these markings are distinct on most segments but are often absent or faint on segments 1 to 5. In *E. vulgata* the markings are present on all segments and are dark and triangular or elongated, often with a pair of longitudinal stripes.

Habitat and Ecology

Nymphs of this species live in the pools and margins of large rivers where they dig into the substrate to form a tubular burrow. They use their gills to force the water through this burrow and the nymphs feed by filtering or collecting fine particulate organic detritus from the water column. *Ephemera lineata* has a two-year life cycle, although some populations may have an annual life cycle. Adults are usually seen in July.

J	F	M	A	M	J	J	A	S	O	N	D
						■					

Emergence of the adults probably takes place at dusk or dawn on the surface of the water or occasionally on a stick, stone or plant stem partially or entirely out of the water.

Distribution

Ephemera lineata is the rarest of the British Ephemeridae. It is currently known from the River Thames and the River Wye (Monmouthshire) but may occur in other watercourses. A voucher specimen is required for all specimens.

Ephemera vulgata **Linnaeus**
Common name: Drake Mackerel Mayfly

Life size

Male sub-imago

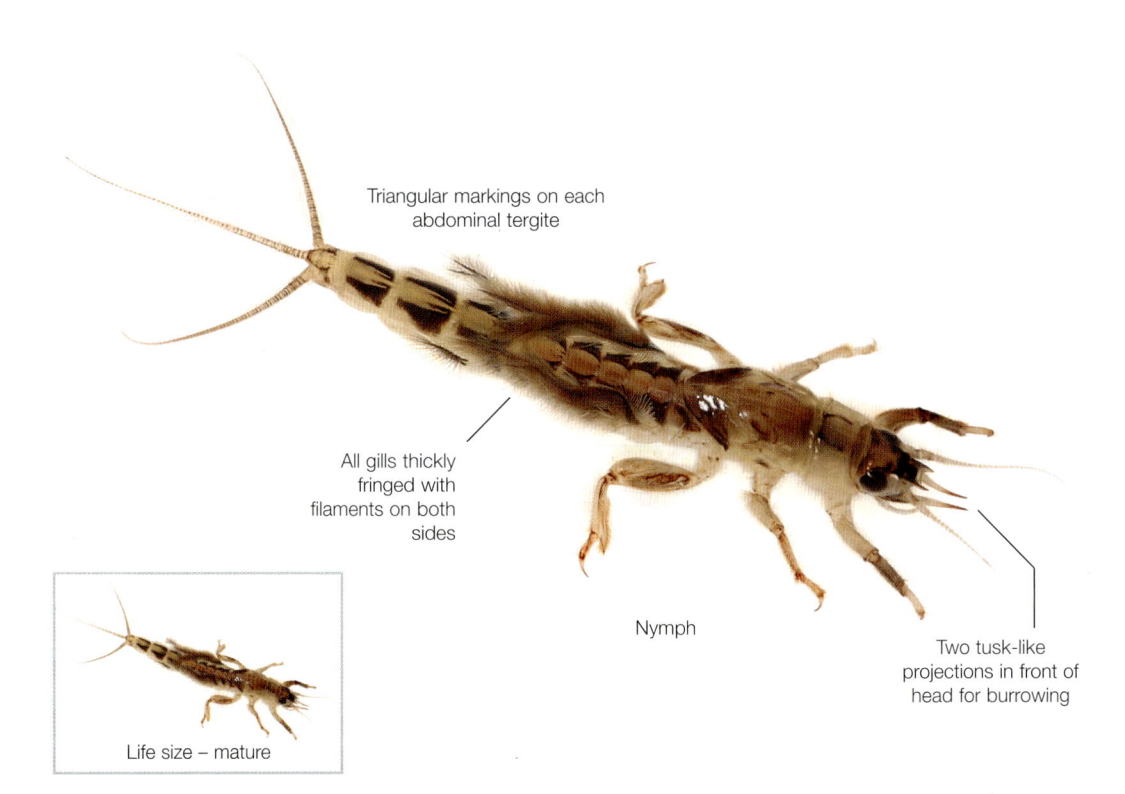

Triangular markings on each abdominal tergite

All gills thickly fringed with filaments on both sides

Nymph

Two tusk-like projections in front of head for burrowing

Life size – mature

Ephemera vulgata Linnaeus
Common name: Drake Mackerel Mayfly

Key features

Nymphs: Large cylindrical nymphs with two tusk-like projections from the head and thick, feathery gills that are held over the back.

Adults: Large flies with three tails and large hindwings. The sub-imago has grey wings, with a yellowish-green tinge and heavy venation. The body is creamy-yellow with distinctive brown markings on the body segments. The tails are dark grey-black and the legs are a creamy-olive colour with black markings. In general, the colours of the wings, tails and legs are darker, and the body paler in the male.

The imago has transparent wings, which have heavy brown venation and several dark patches. The body is creamy-white with the last three segments brownish. The tails are extremely long and almost black in colour.

Separating from other species

Nymphs: The British Ephemeridae can be separated by the markings on the upper surface of their body. The markings in *Ephemera vulgata* are dark and triangular and can be found on all the body segments, although in some cases these are indistinct on the first and last segments. In *E. danica* the markings are neither dark triangles nor a series of lines. On segments 7 to 9 the markings are elongated and obvious, however on the remaining segments the markings are indistinct. In *E. lineata* the markings consist of a series of six stripes on segments 7 to 9. On the remaining segments the markings are indistinct.

Adults: Adults of the British Ephemeridae can also be separated by the markings on the upper surface of their body. In *Ephemera vulgata* the markings are present on all of the segments and are dark and triangular or elongated, often with a pair of longitudinal stripes. In *Ephemera danica* these markings are distinct on most segments but are often absent or faint on segments 1 to 5. *Ephemera lineata* also has markings on each body segment. On segments 5 to 9 these consist of a series of 6 longitudinal stripes, while on other segments they are square, triangular or angular.

Habitat and Ecology

Nymphs of this species live in the pools and margins of muddy rivers where the nymphs dig into the substrate to form a tubular burrow. They use their gills to force the water through this burrow and feed by filtering or collecting fine particulate organic detritus from the water column. *Ephemera vulgata* has a two-year life cycle, although some populations may have an annual life cycle. Adults can be found between May and August.

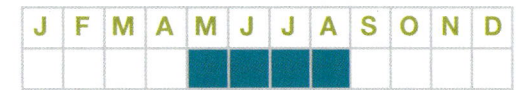

J	F	M	A	M	J	J	A	S	O	N	D
				■	■	■	■				

Emergence of the adults takes place at dusk or dawn on the surface of the water or occasionally on a stick, stone or plant stem partially or entirely out of the water. The males of this species can be found swarming throughout the day, and often swarming continues until dusk.

Once mated, the female flies upstream and descends to the surface of the water to release a few eggs by dipping the tip of her abdomen onto the surface at intervals, or by actually settling on the surface for short periods. After several visits to the water the egg supply of up to 6,000 eggs is finished and the spent female falls on to the surface.

Distribution

Ephemera vulgata is fairly common in the south-east of England and up the east coast as far as Humberside. A voucher specimen is required for specimens from outside these areas.

Identification Chart 5A: **Ephemerellidae – Nymphs**

Ephemerellidae

Nymphs with four pairs of plate-like gills visible. Gills are held above the body and when viewed from above they do not extend beyond the sides of the body. Nymphs swim poorly with a characteristic 'S' shaped rocking motion

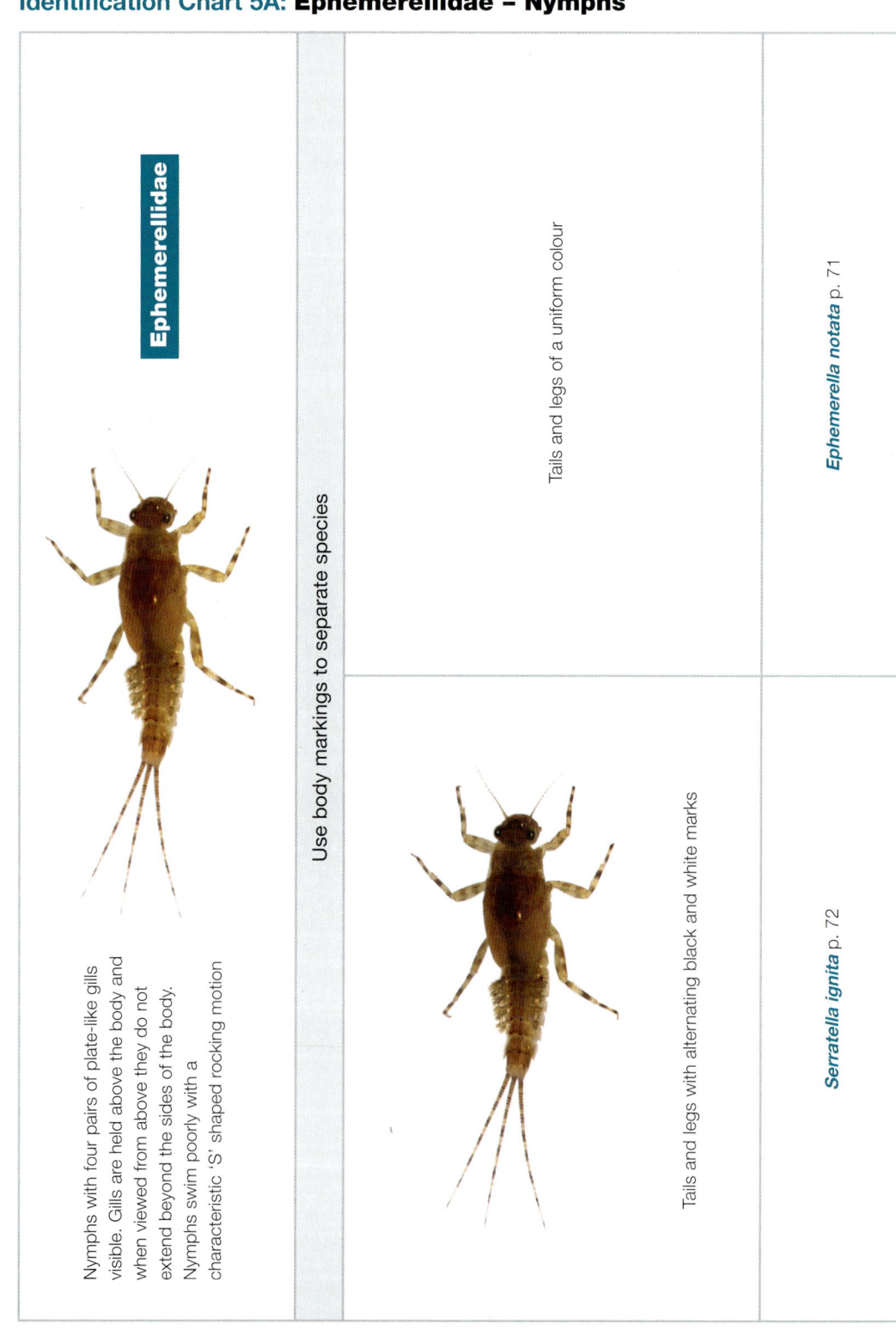

Use body markings to separate species

Tails and legs of a uniform colour

Ephemerella notata p. 71

Tails and legs with alternating black and white marks

Serratella ignita p. 72

Identification Chart 5B: **Ephemerellidae – Adults**

Ephemerellidae

Medium sized fly with regular short veins between each long vein in the forewing.
Forewing size: 7.0-11.0mm

Use body markings to separate species

Underside of body with a pattern of dark lines and dots. Overall colour of imago yellowish

Ephemerella notata p. 71

Underside of body plain with a very faint pattern of lines and dots. Overall colour of imago reddish

Serratella ignita p. 72

Ephemerella notata Eaton
Common name: Yellow Hawk or Yellow Evening Dun

Key features

Nymphs: Nymphs with four pairs of plate like gills visible. Gills are held above the body and when viewed from above they do not extend beyond the sides of the body. Nymphs swim poorly with a characteristic 'S' shaped rocking motion.

Adults: Medium sized flies with three tails and large hindwings. The sub-imago of this species has pale yellowish-grey wings with yellow venation. In the male the body is yellowish, with the last three segments pale amber in colour. In contrast, the female has an orange-yellow body. In both sexes the legs are yellowish and the tails are yellowish with brown rings.

In the imago the wings are transparent with yellowish veins along the leading edges. The body is yellow-olive and the underside has a distinctive pattern of lines and spots. The male has orange coloured eyes, while in the female they are greenish.

Separating from other species

Nymphs: The nymphs of *Ephemerella notata* have a distinctive rocking swimming style. The nymphs are light in colour and generally do not have any patterning on the body, apart from a series of dark marks on the underside of the body. They can be separated from the closely related *Serratella ignita* by the absence of any patterning on the legs and tails. Some workers suggest that the presence of distinctive marks on the underside of *E. notata* nymphs can be used to separate these species. However this is unreliable as the marks are often present, albeit faintly, in *S. ignita*.

A definitive feature is the profile of the body. In *S. ignita* there are a pair of small spines on the upper surface of each body segment. These spines are absent in *E. notata* (see p. 72).

Adults: *Ephemerella notata* is superficially similar to *Habrophlebia fusca*, however in *E. notata* there are small detached single veins at the edges of the forewings, between each of the major veins. *H. fusca* never has detached veins at the edges of the forewings.

Many workers use the presence of distinctive marks on the underside of *E. notata* adults to separate it from *S. ignita*. However this may prove unreliable as the marks are often present, albeit faintly, in *S. ignita*.

Habitat and Ecology

Nymphs of this species live in rivers and streams either on in-stream vegetation or amongst the sand and gravel on the bed. They are usually found clinging to, or crawling amongst submerged plants and stones, although they may swim in short bursts if disturbed. The nymphs feed by collecting or gathering fine particulate organic detritus from the sediment.

There is one generation a year, which can overwinter either as eggs or as nymphs and emerges between May and June.

J	F	M	A	M	J	J	A	S	O	N	D
				■	■						

Males of this species swarm at dusk and once mated, the female produces an egg mass which she holds under her tail. This egg mass is released as the insect flies above the water.

Distribution

Ephemerella notata is less common than *Serratella ignita*, however its range has spread northwards considerably over the past 25 years.

Serratella ignita (Poda)
Common name: Blue Winged Olive

Life size

Female imago

Tails with alternating light and dark bands

Four pairs of visible gills (small 5th pair hidden under 4th)

First three gills with inner rear margin drawn out into an extension

Upper surface of each abdominal segment has two distinct spines on either side of the mid line

Nymph

Life size – mature

Serratella ignita (Poda)
Common name: Blue Winged Olive

Key features

Nymphs: Nymphs with four pairs of plate like gills visible. Gills are held above the body and when viewed from above they do not extend beyond the sides of the body. Nymphs swim poorly with a characteristic 'S' shaped rocking motion.

Adults: Small to medium sized flies with three tails and large hindwings. The sub-imago has dark blue-grey wings and the tails are pale grey and are ringed with brown. Females have dark green eyes and the body is a greenish-olive. As the flight period progresses the body darkens to a rusty brownish-olive. The male has red eyes and the body varies between orange-brown and olive-brown. The last segment is often a much lighter shade.

The imago has transparent wings with pale brown venation. In the female the eyes are greenish-brown while the male has bright red eyes. The body varies between olive-brown and a deep sherry-red, however in the male the body is generally lighter.

Separating from other species

Nymphs: The nymphs of *Serratella ignita* have a distinctive rocking swimming style and have a marked black and white pattern to the legs and tails. The nymphs are generally dark, however this tends to vary with the colour of the substrate. They can be separated from the closely related *Ephemerella notata* by the alternating pattern on the legs and tails. Some workers suggest that the presence of distinctive marks on the underside of *E. notata* can be used to separate these species however this is unreliable as the marks are often present, albeit faintly, in *S. ignita*. A definitive feature is the profile of the body. In *S. ignita* there are a pair of small spines on the upper surface of each body segment. These spines are absent in *E. notata*.

Adults: *Serratella ignita* is superficially similar to *Habrophlebia fusca*, however in *S. ignita* there are small detached single veins at the edges of the forewings, between each of the major veins. *H. fusca* never has detached veins at the edges of the forewings.

Many workers use the presence of distinctive marks on the underside of *E. notata* to separate it from *S. ignita*, however this may prove unreliable as the marks are often present, albeit faintly, in *S. ignita*.

Habitat and Ecology

Nymphs of this species live in fast flowing streams and rivers, especially where aquatic vegetation is present. Occasionally it has been found on the stony shores of upland lakes. They are usually found clinging to, or crawling amongst submerged plants and stones, although they may swim in short bursts if disturbed. The nymphs feed by collecting or gathering fine particulate organic detritus from the sediment.

There is one generation a year, which usually overwinters in the egg stage and emerges between April and September. There may be separate winter and summer generations in warmer waters, such as those of southern England, and this would result in a longer flight period.

J	F	M	A	M	J	J	A	S	O	N	D
			■	■	■	■	■	■			

Emergence of the adults takes place at the surface of the water during daylight hours and at dusk. The males of this species can be found swarming throughout the day, and swarming continues until dusk.

The mated female produces an egg mass which she holds under her tail. The eggs are usually laid in areas of fast flowing and turbulent water, where moss is present. The female flies over the water and releases the egg mass.

Distribution

Serratella ignita is one of the most common and widespread Ephemeroptera species. It can be found throughout the British Isles.

Identification Chart 6A: **Heptageniidae – Nymphs**

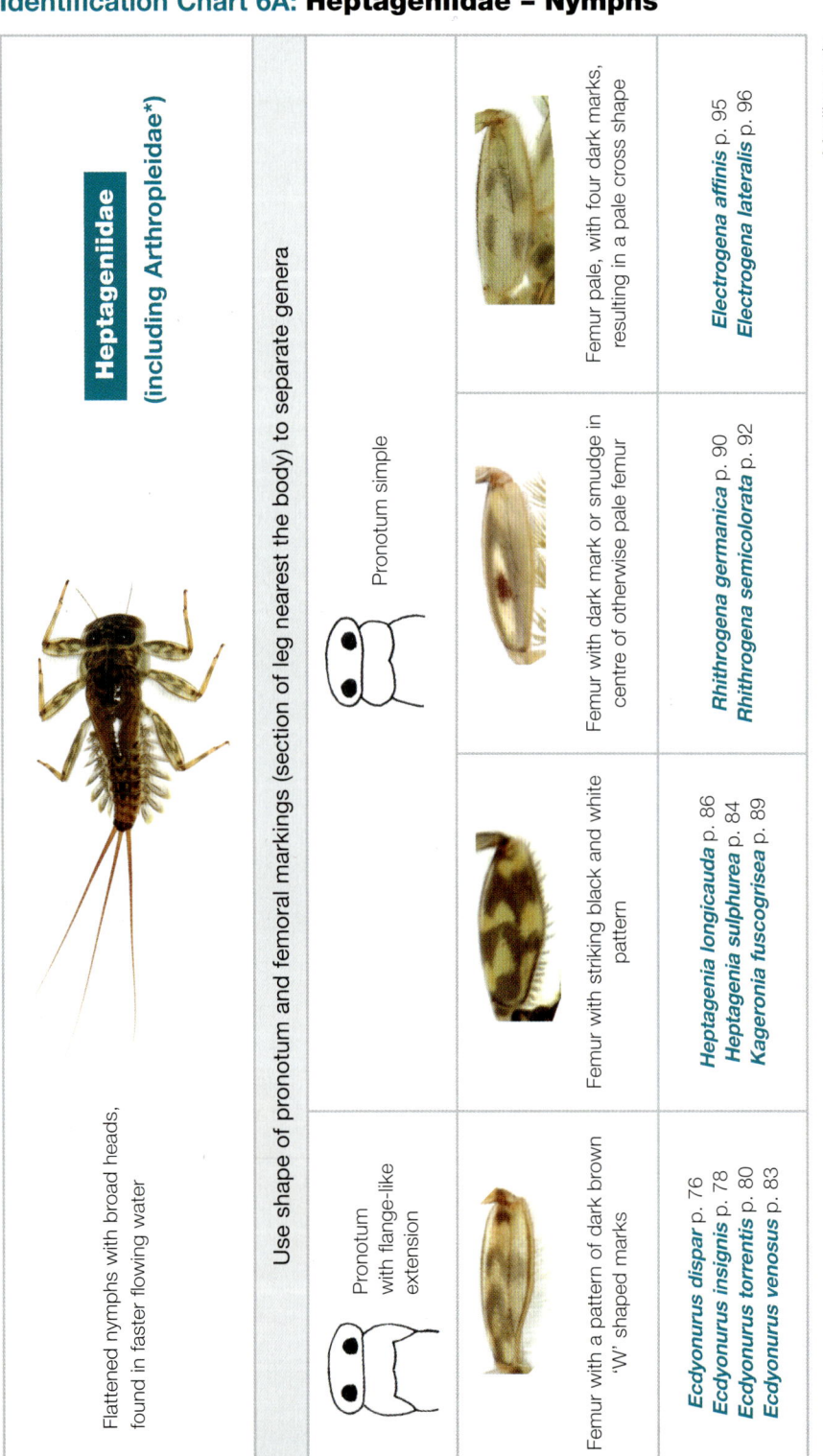

Heptageniidae (including Arthropleidae*)

Flattened nymphs with broad heads, found in faster flowing water

Use shape of pronotum and femoral markings (section of leg nearest the body) to separate genera

Pronotum with flange-like extension	Pronotum simple	
Femur with a pattern of dark brown 'W' shaped marks	Femur with striking black and white pattern	Femur with dark mark or smudge in centre of otherwise pale femur	Femur pale, with four dark marks, resulting in a pale cross shape
Ecdyonurus dispar p. 76 *Ecdyonurus insignis* p. 78 *Ecdyonurus torrentis* p. 80 *Ecdyonurus venosus* p. 83	*Heptagenia longicauda* p. 86 *Heptagenia sulphurea* p. 84 *Kageronia fuscogrisea* p. 89	*Rhithrogena germanica* p. 90 *Rhithrogena semicolorata* p. 92	*Electrogena affinis* p. 95 *Electrogena lateralis* p. 96

Maxillary palp

* *Arthroplea congener* (p. 99) has only been recorded from the British Isles on one occasion. A single adult male was collected by R. South from Stanmore, Middlesex on the 4th June 1920. No further records of this species have been made, and subsequently this species may not be included on future checklists of the British Ephemeroptera. *A. congener* is distinguished immediately by its unique maxillary palp (see figure) and the lack of a tuft of filaments on all the gills.

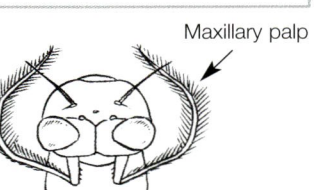

Identification Chart 6B: **Heptageniidae – Adults**

Heptageniidae (including Arthropleidae*)

Small to medium sized flies with two tails and large hindwings. Five free segments in the foot.
Forewing size: 8.0–17.0mm

Examine eyes, legs and feet to separate genera
(The separation of Heptageniidae genera is difficult in all but male imagines.
It is recommended that a specialist key is used where a correct identification is particularly important)

Eyes either touching or very close together		Eyes further apart, never touching		
Femur (section of leg nearest the body) with dark mark or smudge in centre of otherwise pale femur. Body segments plain	No dark mark or smudge in the centre of the femur. Body segments with dark triangular marks at the sides	First segment of the foot (closest to the body) is equal to, or shorter than, segment two		First segment of the hind foot is longer than segment two
		First segment of the foot shorter than segment two	First segment of the foot the same length as segment two	
Rhithrogena germanica p. 90 *Rhithrogena semicolorata* p. 92	*Ecdyonurus dispar* p. 76 *Ecdyonurus insignis* p. 78 *Ecdyonurus torrentis* p. 80 *Ecdyonurus venosus* p. 83	*Heptagenia* *H. longicauda* p. 86 *H. sulphurea* p. 84	*Kageronia* *K. fuscogrisea* p. 89	*Electrogena affinis* p. 95 *Electrogena lateralis* p. 96

* *Arthroplea congener* has only been recorded from the British Isles on one occasion. A single adult male was collected by R. South from Stanmore, Middlesex on the 4th June 1920. This record is now described as doubtful by many writers, and subsequently this species may not be included on future checklists of the British Ephemeroptera. In male imagines of *A. congener*, segments 1 to 4 of the foot in the first pair of legs are nearly equal in length. Each of these segments is about twice the length of the fifth segment.

Ecdyonurus dispar (Curtis)
Common name: Autumn Dun

Life size

Male imago

Gills plate-like and positioned on the side of the abdomen. All but last gill consist of a plate plus a bunch of fine filaments

No tuft of filaments on last gill. First and last gills lying beside the abdomen and not touching under the body

Nymph

Pronotum extending backwards over part of next segment

Life size – mature

Ecdyonurus dispar (Curtis)
Common name: Autumn Dun

Key features

Nymphs: Flattened nymphs with broad heads. The eyes are large and placed on the back of the head. Typically found in faster flowing water.

Adults: Medium sized flies with two tails and large hindwings. Sub-imagines of *Ecdyonurus dispar* have grey wings with a yellowish tinge to the wing membrane and dark brown veins. Female sub-imagines have a brownish-olive body that is decorated with chestnut-brown diagonal markings. These markings are not as prominent in the male, which has a more reddish-brown body. The forelegs in both the male and the female are particularly long. The eyes of the female are dark brown whilst in the male they are greenish-brown. The imagines of both sexes are very similar. Both have transparent wings with prominent dark brown veining. The colour of the eyes is similar to that of the sub-imago, however the body of the female is red-brown, while the male is bright red-brown with blackish edges to the segments.

Separating from other species

Nymphs: The British *Ecdyonurus* are represented by four similar species. The features used to separate these species are either subjective or require microscopic examination.

There are a number of features that may be used with less confidence. Fully mature nymphs of *E. dispar* typically have two dark bands on the foot – one at either end (a). In addition the underside of the body is marked with a dark line along the front edge of each segment (b). A specialist key should be consulted to confirm the identification.

(a)   (b)

Adults: *Ecdyonurus dispar* can be separated from *E. insignis* by examining the underside of the body. In *E. insignis* there is a pattern of dark lines and dots on a yellowish background on each body segment. In other *Ecdyonurus* species the underside is usually plain and reddish in colour.

E. dispar can be separated from other *Ecdyonurus* species by the colour and patterning of the forewing in the sub-imago. In *E. dispar* the forewings are greyish yellow with no obvious patterning and plain, unmarked cross-veins.

The forewings of *E. torrentis* are mottled with some of the cross-veins marked with black. This results in several dark bands beings present across the wing. In *E. venosus* there are black borders to all the cross-veins resulting in a uniform patterning of the wing, with no obvious bands. Identification of the imago is difficult and unreliable.

Habitat and Ecology

Ecdyonurus dispar is typically found in riffle areas of rivers and larger streams. It can also occasionally be found on the wave-lashed shores of larger standing waters. The nymphs are usually found clinging to submerged plants and stones, although they may swim if disturbed. *E. dispar* feeds either by scraping algae from the substrate or by gathering fine particulate organic detritus from the sediment.

There is one generation a year, which usually overwinters as nymphs and emerges between June and October. Males may be found swarming throughout the afternoon.

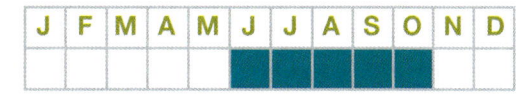

J	F	M	A	M	J	J	A	S	O	N	D
					■	■	■	■	■		

Once mated, the female rests on a stone above the water and lays around 2,500 eggs directly on the river bed by dipping the tip of her abdomen under the surface.

Distribution

Due to the problems with identification of nymphs, it is difficult to compile a complete distribution for this species. *E. dispar* is however, thought to occur in suitable habitats throughout the UK and Ireland.

Ecdyonurus insignis (Eaton)
Common name: Large Green Dun

Life size

Markings on the underside of the body

Male sub-imago

Dark markings on the underside of the body

Tuft of filaments alongside each gill including the last

Nymph

Life size – mature

Ecdyonurus insignis (Eaton)
Common name: Large Green Dun

Key features

Nymphs: Flattened nymphs with broad heads. The eyes are large and placed on the back of the head. The gills consist of a flat plate and a tuft of filaments. Typically found in faster flowing water.

Adults: Large flies with two tails and large hindwings. The sub-imago of *Ecdyonurus insignis* has mottled fawn wings but unlike *Ecdyonurus torrentis*, there is no black mottling but a clear area is present in the centre of the wing. The body of this insect is dark olive-green. The imagines have a grey-black patch along the top of the leading edge of the forewings and the body is olive-green with brown banding.

Separating from other species

Nymphs: Nymphs: *Ecdyonurus insignis* can be separated from other British *Ecdyonurus* species by the presence of dark markings on the underside of the body and the presence of a tuft of filaments alongside the last gill.

Adults: *Ecdyonurus insignis* can be separated from other *Ecdyonurus* species by the presence of a pattern of dark lines and dots on a yellowish background on each body segment. In other *Ecdyonurus* species the underside is usually plain and reddish in colour.

Habitat and Ecology

Nymphs of this species are typically found in riffle areas of rivers and streams. They are usually found clinging to submerged plants and stones, although they may swim if disturbed. *E. insignis* feeds either by scraping algae from the substrate or by gathering fine particulate organic detritus from the sediment.

There is one generation a year, which usually overwinters as nymphs and emerges between May and October.

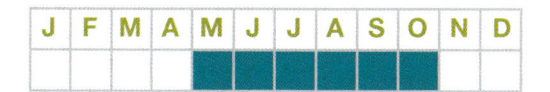

Distribution

Ecdyonurus insignis is a highly localised species with records from a small number of watercourses in southern England and Scotland. This apparent rarity may be due to under-recording. A voucher specimen is required for all records of this species.

Ecdyonurus torrentis **Kimmins**
Common name: Large Brook Dun

Life size

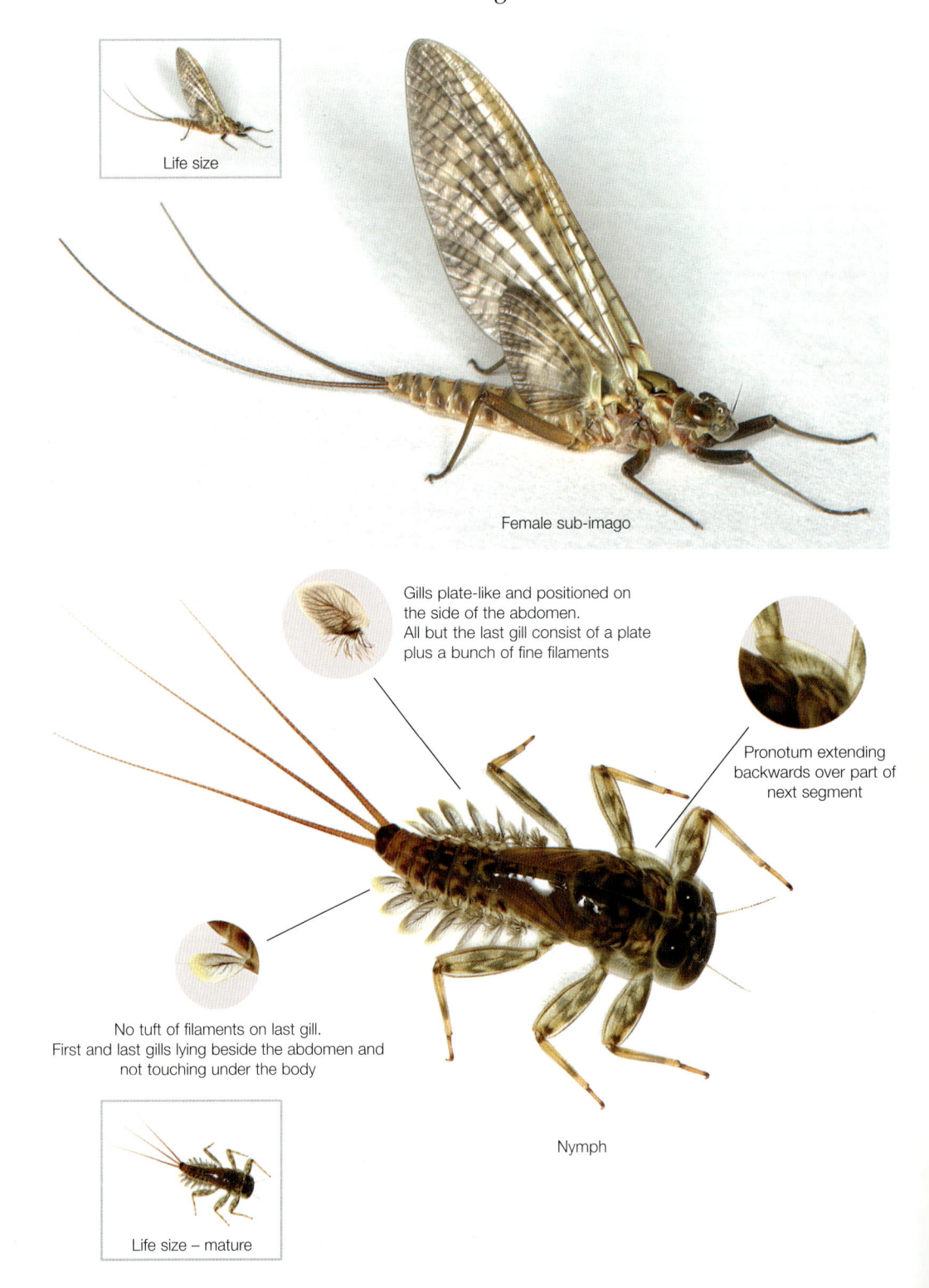

Female sub-imago

Gills plate-like and positioned on the side of the abdomen. All but the last gill consist of a plate plus a bunch of fine filaments

Pronotum extending backwards over part of next segment

No tuft of filaments on last gill. First and last gills lying beside the abdomen and not touching under the body

Nymph

Life size – mature

Ecdyonurus torrentis Kimmins
Common name: Large Brook Dun

Key features

Nymphs: Flattened nymphs with broad heads. The eyes are large and placed on the back of the head. The gills consist of a flat plate together with a tuft of filaments. Typically found in faster flowing water.

Adults: Large flies with two tails and large hindwings. The sub-imago of *Ecdyonurus torrentis* is one of the most distinctive adult mayflies found in the British Isles. The wings are a pale fawn colour, mottled with black patches and with a strong yellow colour on the leading edges. The body of both the male and female is olive-brown with reddish sides which gives a banded appearance to the body. The underbody is purple. Imagines have transparent wings with very dark veining and a yellowish tinge to the leading edge. The body is dark olive-brown with bands of red or purple and, like the sub-imago, the underbody is purple. The female is much larger than the male.

Separating from other species

Nymphs: The British *Ecdyonurus* are represented by four similar species. The features used to separate these species are either subjective or require microscopic examination.

There are a number of features that may be used with less confidence. Fully mature nymphs of *E. torrentis* typically have two dark bands on the foot – one at either end (a). In addition the underside of each body segment is marked with a pattern or pale lines and dots on a dark background (b). A specialist key should be consulted to confirm the identification.

(a)   (b)

Adults: *Ecdyonurus torrentis* can be separated from *E. insignis* by examining the underside of the body. In *E. insignis* there is a pattern of dark lines and dots on a yellowish background on each body segment. In other *Ecdyonurus* species the underside is usually plain and reddish in colour. *E. torrentis* can be separated from other *Ecdyonurus* species by the colour and patterning of the forewing in the sub-imago. In *E. torrentis* the forewings are mottled with some of the cross-veins marked with black. This results in several dark bands being present across the wing. The forewings of *E. dispar* are greyish yellow with no obvious patterning and plain, unmarked cross-veins while in *E. venosus* there are black borders to all the cross-veins resulting in a uniform patterning of the wing, with no obvious bands. Identification of the imago is difficult and unreliable.

Habitat and Ecology

Nymphs of this species are typically found in riffle areas of rivers and streams. They are usually found clinging to submerged plants and stones, although they may swim if disturbed. *E. torrentis* feeds either by scraping algae from the substrate or by gathering fine particulate organic detritus from the sediment.

There is one generation a year, which usually overwinters as nymphs and emerges between May and September, although in some years adults can be found as early as March. The flight period is often related to the altitude of the emergence site. In upstream reaches the flight period can last for up to three months, whereas in the lower reaches the flight period may be as short as one month.

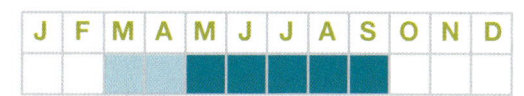

J	F	M	A	M	J	J	A	S	O	N	D
			░	▓	▓	▓	▓	▓			

Once mated, the female flies upstream and descends to the surface of the water to release a few eggs by dipping the tip of her abdomen on to the surface at intervals, or by actually settling on the surface for short periods. After several visits to the water the egg supply of up to 8,300 eggs is finished and the spent female falls on to the surface.

Distribution

Due to the problems with identification of nymphs, it is difficult to compile a complete distribution for this species. *E. torrentis* is however, thought to occur in suitable habitats throughout the UK and Ireland.

Ecdyonurus venosus (Fabricius)
Common name: False or Late March Brown

Key features

Nymphs: Flattened nymphs with broad heads. The eyes are large and placed on the back of the head. The gills consist of a flat plate together with a tuft of filaments. Typically found in faster flowing water.

Adults: Large flies with two tails and large hindwings. The sub-imago of *Ecdyonurus venosus* has fawn coloured, mottled wings. It is similar to *Rhithrogena germanica* but the wing venation is lighter and the body darker. The imago is also similar to that of *R. germanica*, however it lacks the straw-coloured rings around the segments of the abdomen which is a deep mahogany-red, particularly in the female.

Separating from other species

Nymphs: The British *Ecdyonurus* are represented by four similar species. The features used to separate these species are either subjective or require microscopic examination.

There are a number of features that may be used with less confidence. Fully mature nymphs of *E. venosus* typically only have a single dark band on the foot (a) – compare with *E. dispar* (p. 76) and *E. torrentis* (p. 80). The underside of the body is generally unmarked. A specialist key should be consulted to confirm the identification.

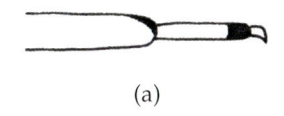

(a)

Adults: *Ecdyonurus venosus* can be separated from *E. insignis* by examining the underside of the body. In *E. insignis* there is a pattern of dark lines and dots on a yellowish background on each body segment. In other *Ecdyonurus* species the underside is usually plain and reddish in colour. *E. venosus* can be separated from other *Ecdyonurus* species by the colour and patterning of the forewing in the sub-imago. In *E. venosus* there are black borders to all the cross-veins resulting in a uniform patterning of the wing, with no obvious bands. The forewings of *E. torrentis* are mottled with some of the cross-veins marked with black. This results in several dark bands being present across the wing. In *E. dispar* the forewings are greyish yellow with no obvious patterning and plain, unmarked cross-veins. Identification of the imago is difficult and unreliable.

Habitat and Ecology

Nymphs of this species are typically found in riffle areas of rivers and streams. They are usually found clinging to submerged plants and stones, although they may swim if disturbed. *E. venosus* feeds either by scraping algae from the substrate or by gathering fine particulate organic detritus from the sediment. There is one generation a year, which usually overwinters as nymphs and emerges between April and July, and often as late as September or October.

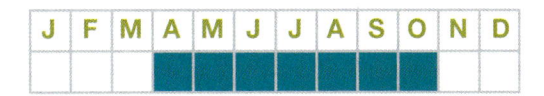

J	F	M	A	M	J	J	A	S	O	N	D
			■	■	■	■	■	■	■		

Emergence of the adults takes place at the surface of the water and often continues well into dusk. The males of this species can be found swarming throughout the day. Once mated, the female either flies upstream and descends to the surface of the water to release a few eggs, by dipping the tip of her abdomen on to the surface at intervals, or rests on a stone above the water and lays her eggs on the river bed by dipping her abdomen below the surface. After several visits to the water the egg supply of up to 6,000 eggs is finished and the spent female falls on to the surface.

Distribution

Due to the problems with identification of nymphs, it is difficult to compile a complete distribution for this species. *E. venosus* is, however, thought to occur in suitable habitats throughout the UK and Ireland.

Heptagenia sulphurea (Müller)
Common name: Yellow May Dun

Life size

Colour of eyes in newly emerged sub-imago

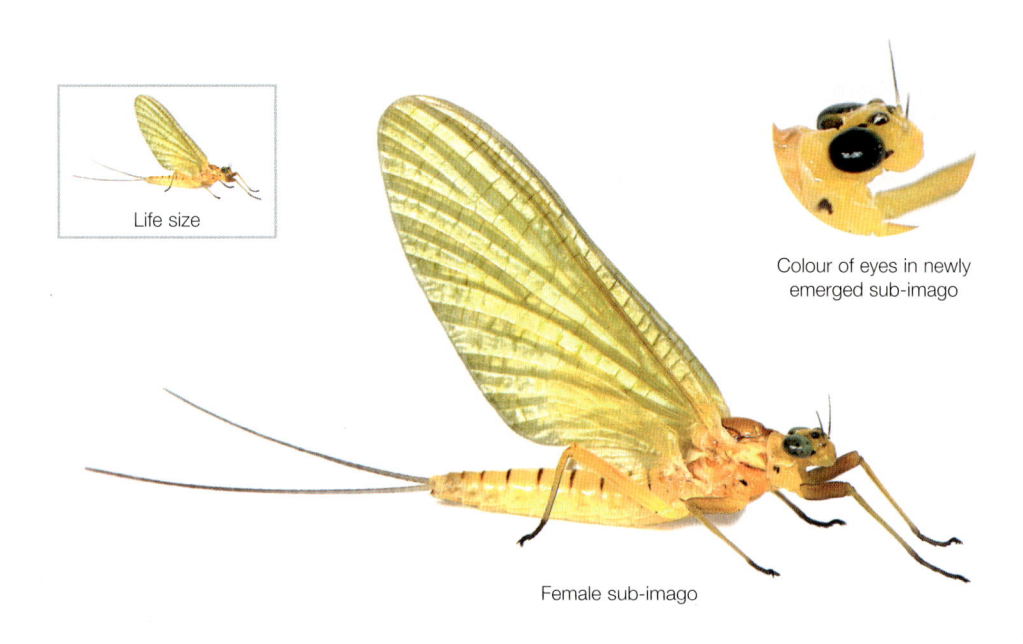

Female sub-imago

Gills plate-like and positioned on the side of the abdomen. All gills consist of a plate plus a bunch of fine filaments. 1st and last gills lie beside the abdomen

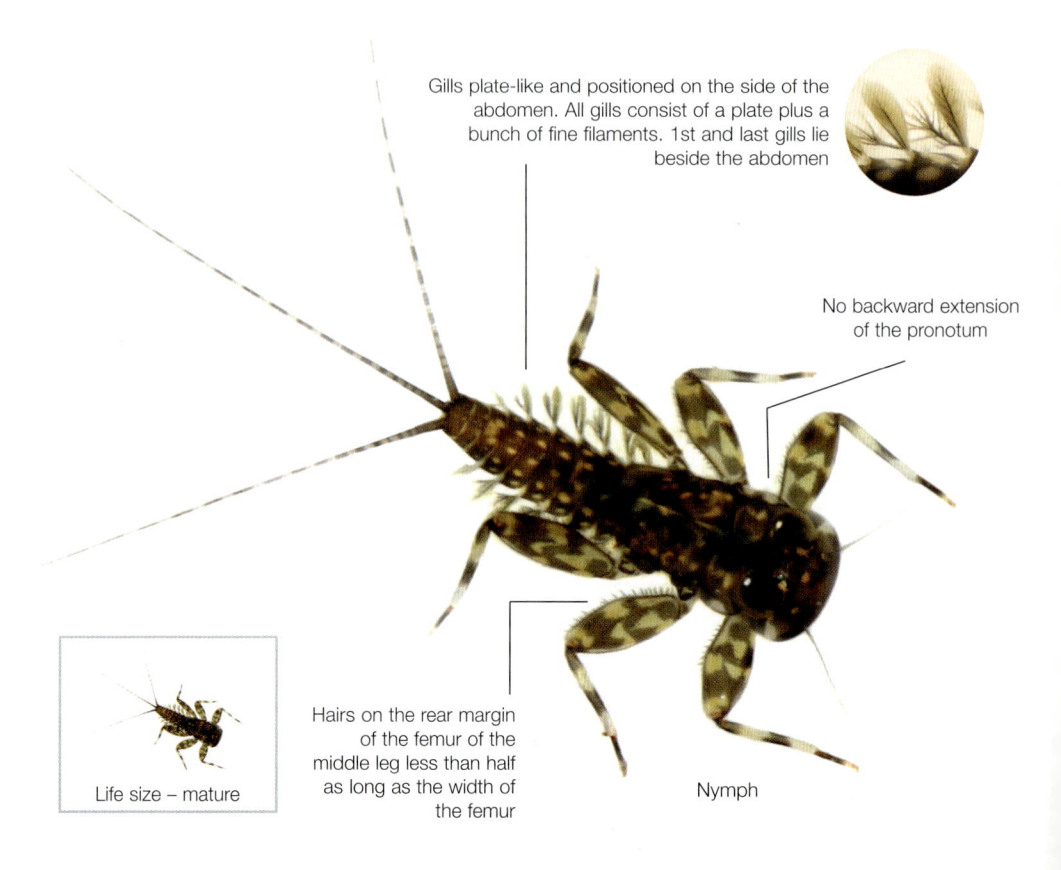

No backward extension of the pronotum

Life size – mature

Hairs on the rear margin of the femur of the middle leg less than half as long as the width of the femur

Nymph

Heptagenia sulphurea (Müller)
Common name: Yellow May Dun

Key features

Nymphs: Flattened nymphs with broad heads. The eyes are large and placed on the back of the head. The gills consist of a flat plate together with a tuft of filaments. Typically found in faster flowing water.

Adults: Medium to large sized flies with two tails and large hindwings. The wings, body and legs of the sub-imago are, as the scientific name suggests, all bright sulphur-yellow. In addition, both the sub-imago and the imago have very distinctive blue eyes. The male imago has a bright golden-brown body, dark brown wing venation and a grey leading edge to the forewings. The female imago is similar but has a pale yellow leading edge to the forewings.

Separating from other species

Nymphs: *H. sulphurea* nymphs are separated from other *Heptagenia* species by the form of the gills and also by the conspicuous black and white pattern on the body. Whilst the current key to the nymphs of the British Ephemeroptera (Elliott, et al., 2010) makes clearer the distinction between *H. sulphurea* and *H. longicauda*, the distinguishing feature used, the filamentous part of the gills, is often damaged during collection. *H. sulphurea* can also be separated from *H. longicauda* by the alternating dark-light pattern on the tails and the absence of small protuberances on the pronotum.

Kageronia fuscogrisea can be separated from *H. sulphurea* by the shape of its gills, which are produced into a sharp point in *K. fuscogrisea*.

Heptagenia species can be seperated from *Electrogena* species by the absence of a long fringe of hairs on the femur (section closest to the body) of each leg.

Adults: *Heptagenia sulphurea* is a very distinctive fly. The sub-imago is bright yellow, a characteristic which is shared with only two other species: *Potamanthus luteus* and *Heptagenia longicauda*.

H. sulphurea can be separated from *P. luteus* by the number of tails present. In *H. sulphurea* there are only two tails while in *P. luteus* there are three. *H. sulphurea* and *H. longicauda* are superficially similar however *H. longicauda* has a pair of flesh-coloured rings on the femur of the front leg, and a single black dot on the side of the body above the hind leg. In *H. sulphurea* there is often up to three black dots above the middle leg, however there are never any dots above the hind leg.

Habitat and Ecology

Heptagenia sulphurea lives chiefly in the riffle sections of larger rivers although it has also been found along the wave lashed shores of calcareous lakes. It feeds either by scraping algae from the substrate or by gathering fine particulate organic detritus from the sediment. The nymphs usually swim in short bursts, interspersed with periods of clinging to submerged plants and stones. There is one generation per year which often has a group of fast growing individuals which emerge en masse in May or June and a slower growing group that emerges in trickles between June and September.

J	F	M	A	M	J	J	A	S	O	N	D
				■	■	■	■	■			

Emergence of the adults typically takes place during daylight or at dusk and the males swarm from afternoon to dusk. Once mated, the female flies upstream and descends to the surface of the water to release a few eggs by dipping the tip of her abdomen on to the surface of the water at intervals, or by actually settling on the surface for short periods. After several visits to the water the egg supply is finished and the spent female falls onto the surface.

Distribution

Heptagenia sulphurea is a widespread and common species, which is found throughout the British Isles. A voucher specimen is required for records of this species from the south-east of England where it is relatively rare.

Heptagenia longicauda (Stephens)
Common name: Scarce Yellow May Dun

Key features

Nymphs: Flattened nymphs with broad heads. The eyes are large and placed on the back of the head. The gills consist of a flat plate together with a tuft of filaments. Typically found in faster flowing water amongst instream vegetation.

Adults: Medium sized flies with two tails and large hindwings. The sub-imago resembles that of *Heptagenia sulphurea*. The wings are pale yellow with a slightly darker shade of yellow along the leading edge. The body and legs are also yellow, and the front legs have two light reddish rings on the femur. Imagines are also similar to those of *H. sulphurea* however in *H. longicauda* the body is more yellowish and each body segment has a darker front edge. In the males body segments 2 to 8 are translucent, while segments 9 and 10 are yellowish-brown. The wings are transparent with dark veins and a yellowish leading edge. In common with the sub-imago the front legs have two light reddish rings on the femur.

Separating from other species

Nymphs: *Heptagenia longicauda* nymphs can be separated from *H. sulphurea* by their plain unbanded tails. In addition, *H. longicauda* has conspicuous protuberances on either side of the pronotum.

Adults: *Heptagenia longicauda* is a very distinctive fly. The sub-imago is bright yellow, a characteristic which is shared with only two other species: *Potamanthus luteus* and *Heptagenia sulphurea*.

H. longicauda can be separated from *P. luteus* by the number of tails present. In *H. longicauda* there are only two tails while in *P. luteus* there are three. *H. sulphurea* and *H. longicauda* are superficially similar. However *H. longicauda* has a pair of flesh-coloured rings on the femur of the front leg, and a single black dot on the side of the body above the hind leg. In *H. sulphurea* there is often up to three black dots above the middle leg, however there are never any dots above the hind leg.

Habitat and Ecology

Recent studies (Teufert, 2001) on *Heptagenia longicauda* have shown that the nymphs are found on submerged bankside vegetation in the riffles and shallows of lowland rivers, particularly in areas of low current. As a result, mature nymphs are easy to collect by hand from aquatic vegetation from May onwards. The nymphs feed on algae and organic detritus that they gather from the substrate or scrape from submerged surfaces. It was previously thought that populations of *H. longicauda* and *H. sulphurea* did not co-exist, but it is now understood that these species do occur together, although *H. longicauda* tends to avoid watercourses where there are large populations of *H. sulphurea* (Haybach, A. pers. comm.). It is likely that where these species do occur together, they occupy distinct micro-habitats and competition between the two species is negligible.

In Europe, the flight period of *H. longicauda* is between May and September, although in warm weather, specimens have been collected as late as October. British specimens have been taken in late May and early June.

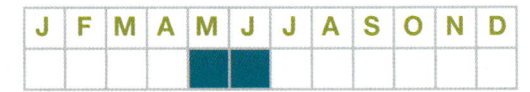

J	F	M	A	M	J	J	A	S	O	N	D
				■	■						

Sub-imagines begin to emerge around sunset and continue to emerge whilst there is any light. In Germany, swarms of imagines have been observed up to 22:30hrs when darkness prevented further observations. It is possible that swarming continues after this time. The imagines form loose swarms of approximately 50 individuals over bridges and other prominent bankside markers. The adults are attracted to a mercury vapour light trap and on calm nights adults have also been observed gathering around bankside lights.

Distribution

Most recent accounts of this species list three British records. In 1868, the Reverend A.E. Eaton collected a sub-imago from near the Kennet and Holybrook by Reading. A female specimen was collected from near the Thames at Staines by E.E. Austen on the 19th May 1904, whilst the most recent specimen, a male sub-imago, was collected on 28th May 1933 by D.E. Kimmins from a young birch tree, close to the River Wey between Tilford and Elstead. There is, however, a further British record. J.F. Stephens first described this species in 1835 from a specimen he collected near Hertford in mid-June. A voucher specimen is required for all records of this species.

Kageronia fuscogrisea (Retzius)
Common name: Brown May Dun

Key features

Nymphs: Flattened nymphs with broad heads. The eyes are large and placed on the back of the head. The gills consist of a flat plate together with a tuft of filaments. Typically found in slower flowing water, often where there is much marginal and emergent vegetation.

Adults: Medium to large flies with two tails and large hindwings. The sub-imago of *Kageronia fuscogrisea* resembles those of *Ecdyonurus venosus* in size and general colour. The body is brownish and the wings are a mottled brownish-fawn colour. The legs are light brown and have two distinct light red bands on each femur.

In the imago the male resembles that of *Heptagenia sulphurea* with a brownish body and dark brown wing venation. The female imago is similar but has a slightly darker body. In both sexes there are two light red bands on each femur.

Separating from other species

Nymphs: *Kageronia fuscogrisea* can be separated from *H. sulphurea* by the shape of its gills, which are produced into a sharp point.

Adults: Sub-imagines are superficially similar to *Ecdyonurus venosus*, but they can be separated by the presence of two light red bands on each femur.

Imagines resemble those of *Heptagenia sulphurea*, but they can also be separated by the presence of two light red bands on each femur.

Habitat and Ecology

Nymphs of *Kageronia fuscogrisea* live chiefly in the riffle sections of larger rivers although they are also found along the wave-lashed shores of calcareous lakes and, atypically for Heptageniidae, amongst stands of macrophytes in standing waters. They feed either by scraping algae from the substrate or by gathering fine particulate organic detritus from the sediment. The nymphs usually swim in short bursts, interspersed with periods of clinging to submerged plants and stones. There is one generation a year, which usually overwinters as nymphs and emerges between May and June. Males swarm throughout the day.

J	F	M	A	M	J	J	A	S	O	N	D
				■	■						

Distribution

Kageronia fuscogrisea is a highly localised species. It is most common in Ireland, however it can also be found in Dumfries and Galloway and occasionally in the catchment of the River Thames. A voucher specimen is required for all records of this species.

Rhithrogena germanica Eaton
Common name: March Brown

Life size

Female sub-imago

Life size – mature

Nymph

Rhithrogena germanica Eaton
Common name: March Brown

Key features

Nymphs: Flattened nymphs with broad heads. The eyes are large and placed on the back of the head. The gills consist of a flat plate together with a tuft of filaments. Typically found in faster flowing water.

Adults: Large flies with two tails and large hindwings. The sub-imagines have dark, mahogany-brown bodies, with the posterior edges of each segment with straw-coloured rings. The wings are strongly mottled with black, however there is a distinctive clear patch that lacks cross-veins in the middle of the wing. The legs are brown to pale brown in colour and the tails are dark brown. The eyes, which are green, have a distinctive dark bar across the centre. The imago has transparent wings with brown venation. The body is similar to the sub-imago, although it is slightly redder. In all stages, including the nymph, there is a dark, oval spot on each femur.

Separating from other species

Nymphs: The British *Rhithrogena* are represented by *R. germanica* and *R. semicolorata*. Unfortunately there are no reliable characters to allow the safe separation of these species. Both species can be found in medium to large watercourses where they crawl amongst the stony substratum. Mature larvae found in late March/early April are likely to be *R. germanica*.

Adults: Adult Heptageniidae have two tails and large hindwings in common with the Siphlonuridae, however in the Heptageniidae there are five free segments in the foot. In the Siphlonuridae there are four free segments.

Rhithrogena germanica is a very distinctive fly, but it is often confused with *Ecdyonurus torrentis* and *E. venosus*. Fortunately it can be readily separated from these species by the presence of a dark mark on the centre of an otherwise pale femur.

Habitat and Ecology

Rhithrogena germanica lives chiefly in the riffle sections of larger rivers or their tributaries, where it feeds either by scraping algae from the substrate or by gathering fine particulate organic detritus from the sediment. The nymphs are usually found clinging to submerged plants and stones, although they may swim if disturbed. There is one generation per year that overwinters as nymphs. This species shows great synchronicity in its emergence with adults hatching en-masse, typically around mid-day, over several days between late March and early May. The sub-imagines immediately fly to the bank where they rest briefly before flying up into trees or other vegetation. They rest there for up to four days before moulting to the imago and returning to the water to swarm and mate.

J	F	M	A	M	J	J	A	S	O	N	D
		■	■	■							

Once mated, the female flies upstream and descends to the surface of the water to release a few eggs by dipping the tip of her abdomen on to the surface at intervals, or by actually settling on the surface for short periods. After several visits to the water the egg supply is finished and the spent female falls on to the surface.

Distribution

Due to the problems with identification, it is difficult to compile a complete distribution for this species. *R. germanica* is however, thought to be the least common of the British *Rhithrogena* species.

Rhithrogena semicolorata (Curtis)
Common name: Olive Upright

Life size

Male sub-imago

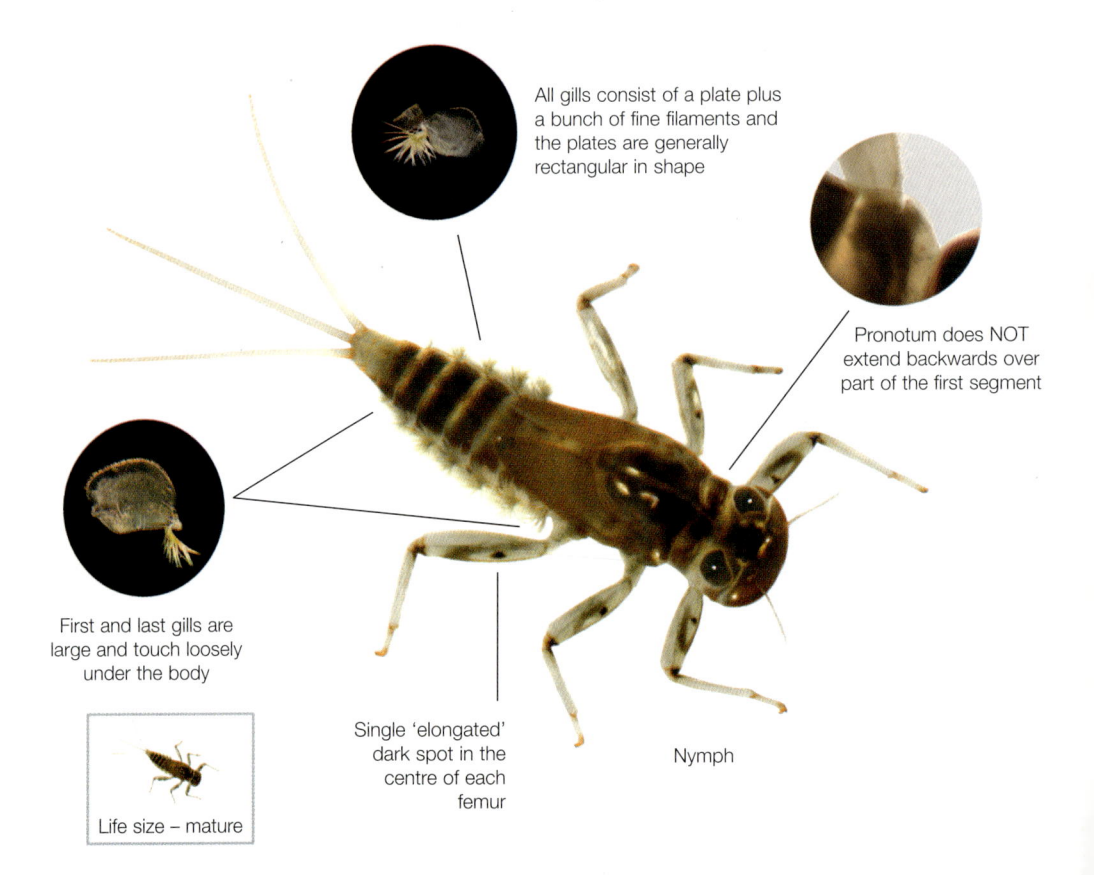

All gills consist of a plate plus a bunch of fine filaments and the plates are generally rectangular in shape

Pronotum does NOT extend backwards over part of the first segment

First and last gills are large and touch loosely under the body

Single 'elongated' dark spot in the centre of each femur

Nymph

Life size – mature

Rhithrogena semicolorata (Curtis)
Common name: Olive Upright

Key features

Nymphs: Flattened nymphs with broad heads. The eyes are large and placed on the back of the head. The gills consist of a flat plate together with a tuft of filaments. Typically found in faster flowing water.

Adults: Medium sized flies with two tails and large hindwings. The sub-imagines have greyish-brown bodies, which are ringed with olive in the male. The side of the thorax is a distinctive orange colour in both the male and female. The forewings are dark blue-grey whilst the hindwings are buff coloured. The wings in the imago have light brown veins and the lower halves have a bronze tinge. The body of both the male and the female varies between brown and light olive.

Separating from other species

Nymphs: The British *Rhithrogena* are represented by *R. germanica* and *R. semicolorata*. Unfortunately there are no reliable characters to allow the safe separation of these species. Both species can be found in medium to large watercourses where they crawl amongst the stony substratum. Mature larvae found in the late spring are likely to be *R. semicolorata*.

Adults: Adult Heptageniidae have two tails and large hindwings in common with the Siphlonuridae, however in the Heptageniidae there are five free segments in the foot. In the Siphlonuridae there are four free segments.

Rhithrogena semicolorata is superficially similar to *Baetis rhodani* and *B. atrebatinus*, but it can be readily separated from these species by its larger hindwings and the presence of a dark mark on the centre of an otherwise pale femur.

Habitat and Ecology

Nymphs of *Rhithrogena semicolorata* live chiefly in the riffle sections of rivers, where they feed either by scraping algae from the substrate or by gathering fine particulate organic detritus from the sediment. The nymphs are usually found clinging to submerged plants and stones, although they may swim if disturbed.

There is one generation per year that overwinters as nymphs. As growth rates vary with water temperature, the period over which adults emerge is variable, but adults have been found between April and September in the British Isles.

J	F	M	A	M	J	J	A	S	O	N	D
		■	■	■	■	■	■	■			

Emergence of the adults typically takes place from dawn until dusk at the surface of the water. Males can be found swarming throughout the day but rarely into the evening. Once mated, the female flies upstream and descends to the surface of the water to release a few eggs by dipping the tip of her abdomen on to the water surface at intervals, or by actually settling on the surface for short periods. After several visits to the water the egg supply is finished and the spent female falls on to the surface.

Distribution

Due to the problems with identification, it is difficult to compile a complete distribution for this species. *R. semicolorata* is, however, thought to be the more common of the British *Rhithrogena* species.

Electrogena affinis (Eaton)
Common name: Scarce Dusky Yellowstreak

Key features

Nymphs: Flattened nymphs with broad heads. The eyes are large and placed on the back of the head. The gills consist of a flat plate together with a tuft of filaments. Typically found in deep, slower flowing water.

Adults: Small to medium flies with two tails and large hindwings. The sub-imago has a yellowish-brown body with reddish markings on each body segment. The wings are grey with a yellowish or sometimes greenish tinge depending on the age of the specimen.

The imago is similar to the sub-imago however the body is lighter and there are still reddish markings on each body segment. In addition, there is a bright yellow streak at the base of the forewings. The wings are milky white, with the wing tip particularly opaque. The cross-veins are indistinct in the male, however they are brownish along the leading edge of the forewing. The legs are yellow with a chestnut-brown stripe extending from the body towards the femur. The tails are whitish with brownish rings on the segments closest to the body. The female imago is similar, however the wings are transparent and the cross-veins are distinctly brownish.

Separating from other species

Nymphs: *Electrogena* species can be separated from other Heptageniidae by the pale cross-shaped pattern on the femur and the presence of a long fringe of hairs on the femur. *E. affinis* was found in the River Derwent, Yorkshire in 1988, and it might be present in other watercourses. The heads of *E. affinis* larvae typically have white patches along the front margin, however, to confirm the identification a specialist key should be consulted.

Adults: *Electrogena* species can be separated from other Heptageniidae by the presence of a bright yellow streak directly below where the forewing is attached to the body.

Electrogena affinis can be distinguished from *E. lateralis* by the colour of the legs and the body. In *E. affinis* the legs are yellowish with a chestnut-brown stripe leading from the body onto the femur. The body is a light creamy-brown colour and has a series of dark chestnut markings on the upper surface. In *E. lateralis* the legs and body are uniformly dark brown.

Habitat and Ecology

Nymphs of this species are typically found in slower flowing areas of lowland rivers, often where there is abundant marginal and emergent vegetation. They are usually found clinging to submerged plants and stones, although they may swim if disturbed. *E. affinis* feeds either by scraping algae from the substrate or by gathering fine particulate organic detritus from the sediment. There is one generation a year, which usually overwinters as nymphs and emerges between July and August.

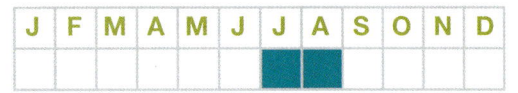

J	F	M	A	M	J	J	A	S	O	N	D
						■	■				

Distribution

E. affinis was found in the River Derwent, Yorkshire in 1988 and is likely to be present in other watercourses. A voucher specimen is required for all records.

Electrogena lateralis (Curtis)
Common name: Dusky Yellowstreak

Life size

Male sub-imago

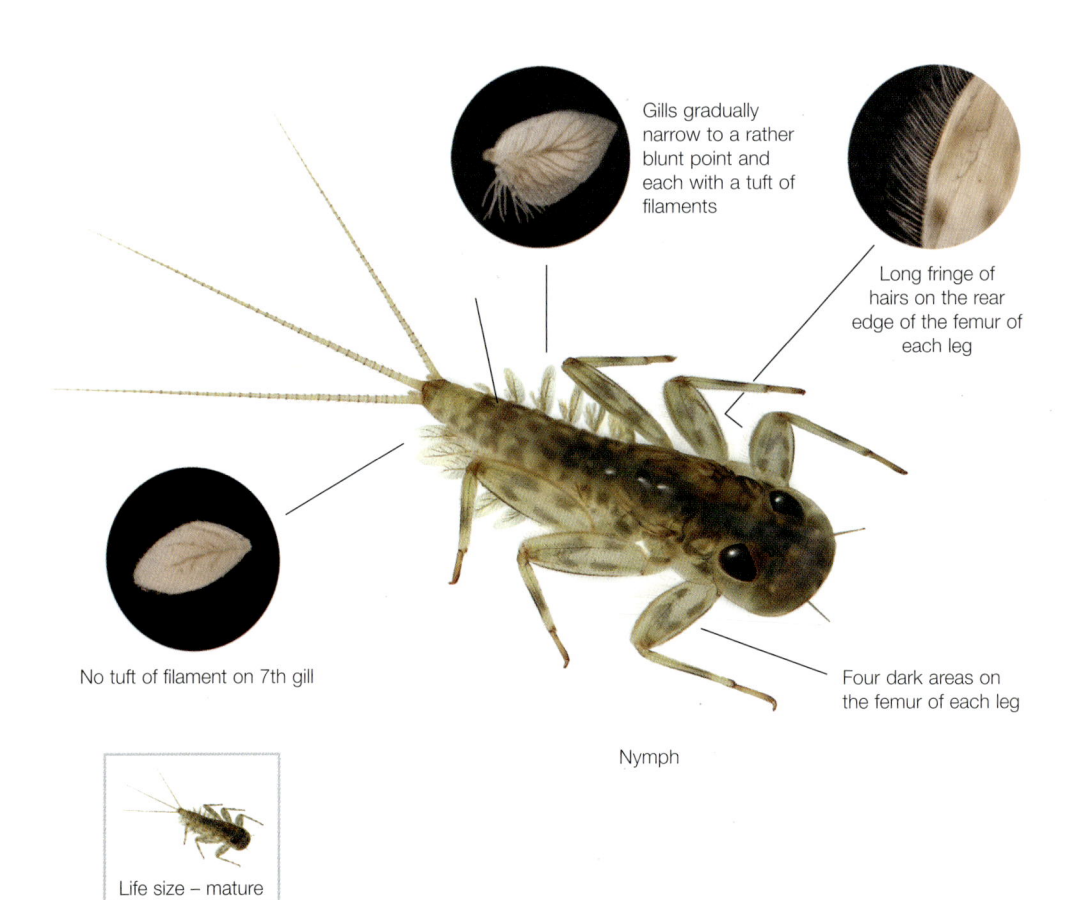

Gills gradually narrow to a rather blunt point and each with a tuft of filaments

Long fringe of hairs on the rear edge of the femur of each leg

No tuft of filament on 7th gill

Four dark areas on the femur of each leg

Nymph

Life size – mature

Electrogena lateralis (Curtis)
Common name: Dusky Yellowstreak

Key features

Nymphs: Flattened nymphs with broad heads. The eyes are large and placed on the back of the head. The gills consist of a flat plate together with a tuft of filaments. Typically found in faster flowing water.

Adults: Medium sized flies with two tails and large hindwings. The sub-imago of *Electrogena lateralis* has wings which are a dark grey colour. The body of both the male and female is dark greyish-brown, as are the tails. Imagines have transparent wings with brownish veins along the leading edge of the forewings. The body is a dark brown-olive colour and each segment of the body is ringed in deep red. The last three body segments of the female are an orange-brown colour, and there is a distinct yellow-orange streak on each side of the thorax directly below each wing root.

Separating from other species

Nymphs: *Electrogena* species can be separated from other Heptageniidae by the pale cross-shaped pattern on the femur and the presence of a long fringe of hairs on the femur. *E. affinis* was found in the River Derwent, Yorkshire in 1988, and it might be present in other watercourses. The head of *E. affinis* larvae typically have white patches along the front margin, however, to confirm the identification a specialist key should be consulted.

Adults: *Electrogena* species can be separated from other Heptageniidae by the presence of a bright yellow streak directly below where the forewing is attached to the body.

Electrogena lateralis can be distinguished from *E. affinis* by the colour of the legs and the body. In *E. lateralis* the legs and body are uniformly dark brown, whereas in *E. affinis* the legs are yellowish with a chestnut-brown stripe leading from the body onto the femur. The body of *E. affinis* is also much lighter and has a series of dark chestnut markings on the upper surface.

Habitat and Ecology

Electrogena lateralis is typically found in riffle areas of rivers and streams, although it can also occasionally be found on the wave-lashed shores of standing waters. The nymphs are usually found clinging to submerged plants and stones, although they may swim if disturbed. *E. lateralis* feeds either by scraping algae from the substrate or by gathering fine particulate organic detritus from the sediment. There is one generation a year, which usually overwinters as nymphs and emerges between May and September.

J	F	M	A	M	J	J	A	S	O	N	D
				■	■	■	■	■			

Whilst this species can emerge at the surface of the water, it is unusual in that it can also emerge under water on the surface of a submerged stone. This allows the adult fly to leave the water surface immediately on reaching the surface thus minimising the possibility of capture by fish or surface-skimming birds such as swallows and sand martins. Once mated, the female flies upstream and descends to the surface of the water to release a few eggs by dipping the tip of her abdomen on to the surface at intervals, or by actually settling on the surface for short periods. After several visits to the water the egg supply of up to 2,500 eggs is finished and the spent female falls on to the surface.

Distribution

Electrogena lateralis is a common, though localised species in Scotland, Wales and the north and south-west of England.

Arthroplea congener Bengtsson

Key features

Nymphs: Flattened nymphs with broad heads. The eyes are large and placed on the back of the head. Long brush-like structures (maxillary palps) extend forward from the head.

Adults: Medium sized flies with two tails and large hindwings. The body is dark brown, almost black with each segment slightly paler on the sides and rear edge. Segments 2 to 8 have two dots on either side of the mid-line. The wings are slightly opaque, particularly in the lower half. The legs are brown to dark brown and the foot is much paler. The tails are a smoky grey colour with rings at the joint of each segment, particularly in the segments closer to the body.

Separating from other species

Nymphs: *Arthroplea congener* has distinctive maxillary palps (see below), which make it unmistakable for any other species.

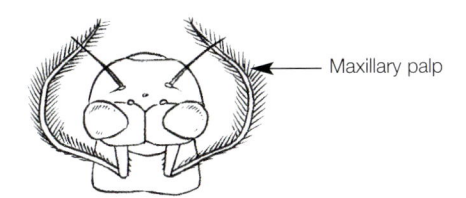

— Maxillary palp

Adults: The separation of *Arthroplea congener* from other Heptageniidae relies upon features found in male imagines. It is recommended that a specialist key is used if you suspect that you may have a specimen of *A. congener*.

Habitat and Ecology

Nymphs of this species live chiefly in the pools and margins of rivers and streams or in small ponds (A. Haybach, pers. comm.), where they cling to submerged plants and stones, although they may swim if disturbed. They feed by filtering of gathering fine particulate organic detritus from the sediment. There is one generation a year, which usually overwinters as eggs and emerges between May and June.

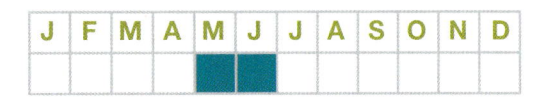

J	F	M	A	M	J	J	A	S	O	N	D
				■	■						

Distribution

Arthroplea congener probably no longer occurs in the British Isles. The only record is of a single adult male taken in 1920 at Stanmore, Middlesex. As a result, a voucher specimen is required for all records of this species.

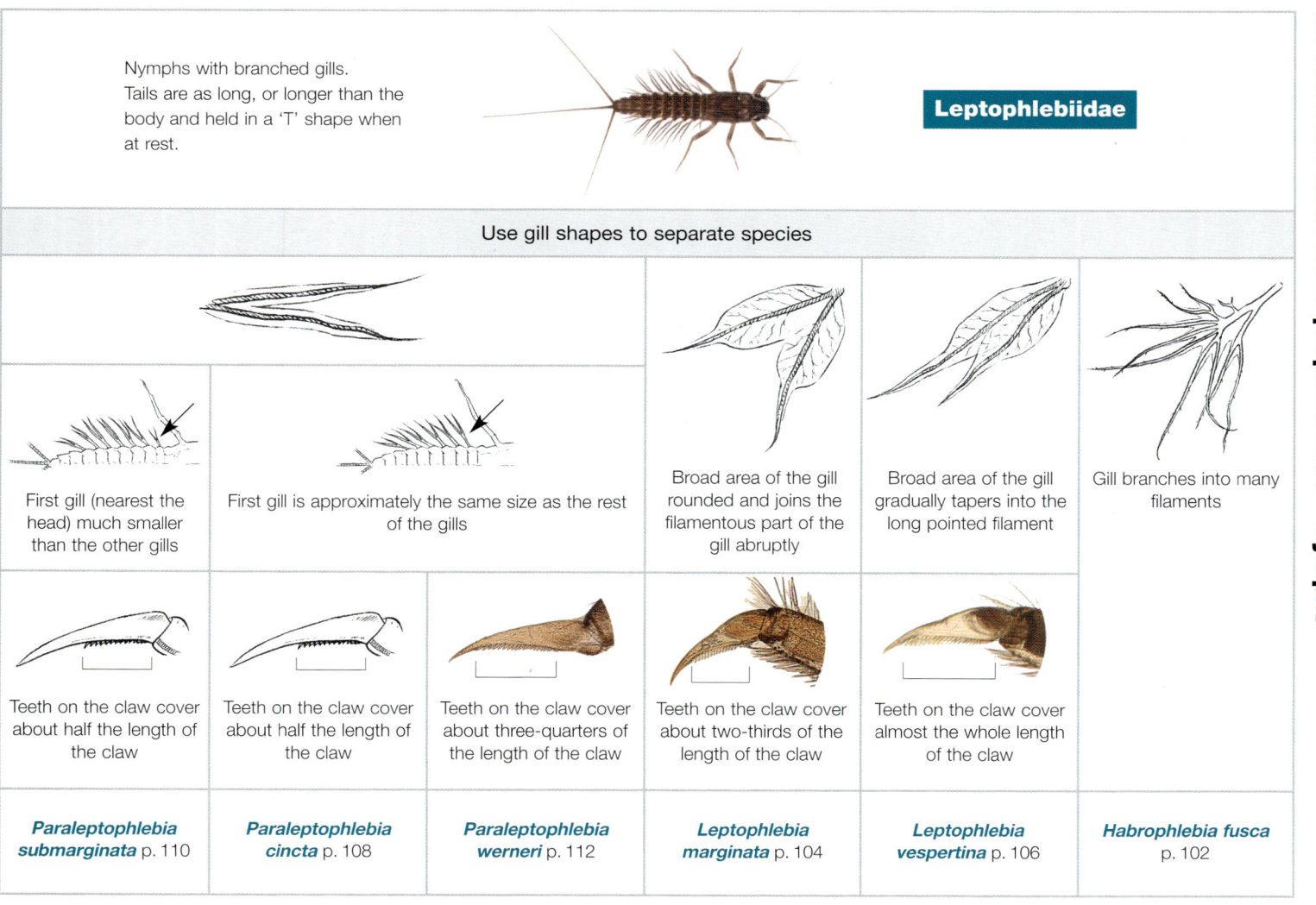

Identification Chart 7A: **Leptophlebiidae – Nymphs**

Leptophlebiidae

Nymphs with branched gills. Tails are as long, or longer than the body and held in a 'T' shape when at rest.

Use gill shapes to separate species

Gill shape	Claw	Species
Gill branches into many filaments		*Habrophlebia fusca* p. 102
Broad area of the gill gradually tapers into the long pointed filament	Teeth on the claw cover almost the whole length of the claw	*Leptophlebia vespertina* p. 106
Broad area of the gill rounded and joins the filamentous part of the gill abruptly	Teeth on the claw cover about two-thirds of the length of the claw	*Leptophlebia marginata* p. 104
First gill is approximately the same size as the rest of the gills	Teeth on the claw cover about three-quarters of the length of the claw	*Paraleptophlebia werneri* p. 112
	Teeth on the claw cover about half the length of the claw	*Paraleptophlebia cincta* p. 108
First gill (nearest the head) much smaller than the other gills	Teeth on the claw cover about half the length of the claw	*Paraleptophlebia submarginata* p. 110

Identification Chart 7B: **Leptophlebiidae – Adults**

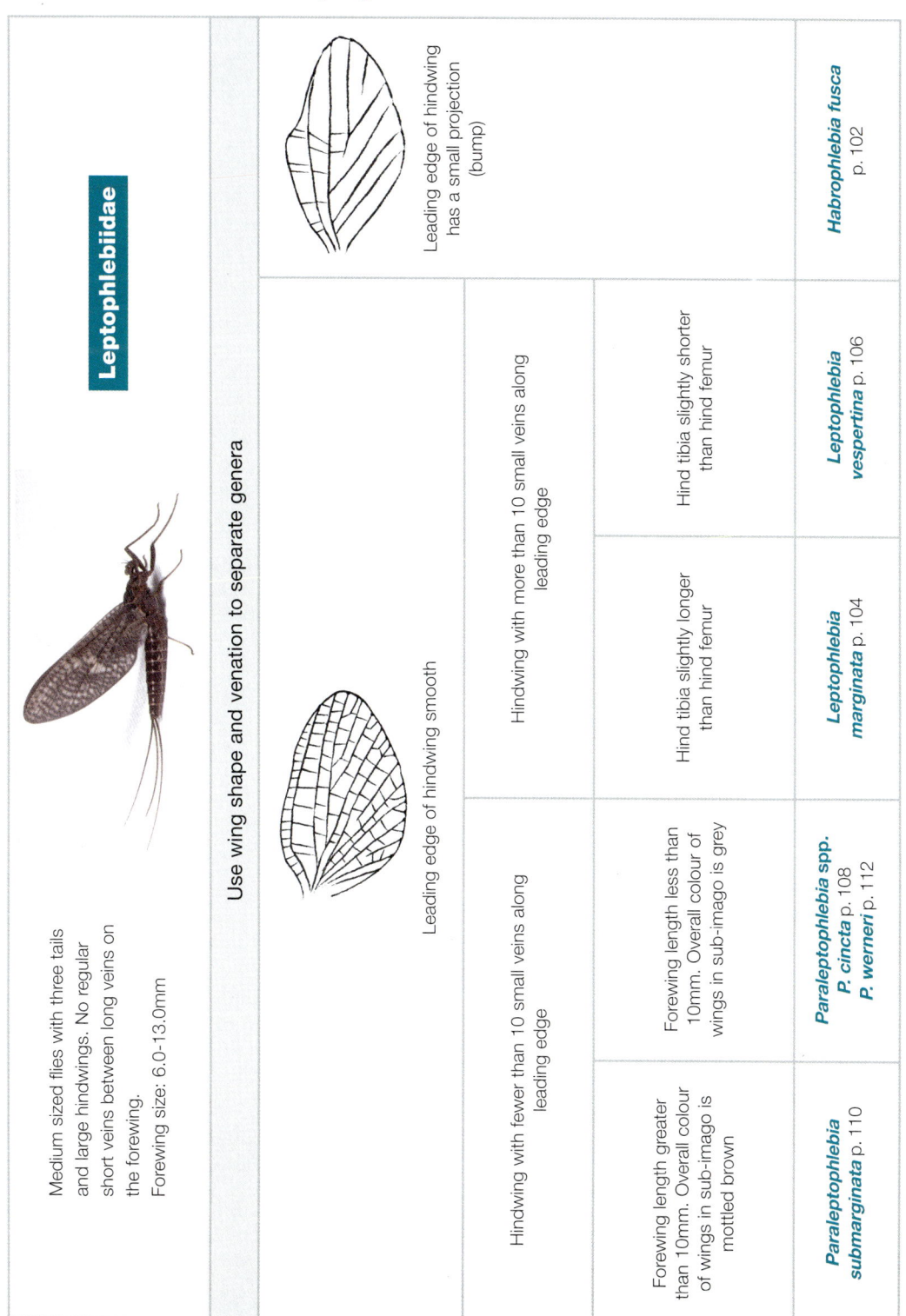

Leptophlebiidae

Medium sized flies with three tails and large hindwings. No regular short veins between long veins on the forewing.
Forewing size: 6.0–13.0mm

Use wing shape and venation to separate genera

Leading edge of hindwing	Further characters	Species
Leading edge of hindwing has a small projection (bump)		*Habrophlebia fusca* p. 102
Leading edge of hindwing smooth — Hindwing with more than 10 small veins along leading edge	Hind tibia slightly shorter than hind femur	*Leptophlebia vespertina* p. 106
	Hind tibia slightly longer than hind femur	*Leptophlebia marginata* p. 104
Leading edge of hindwing smooth — Hindwing with fewer than 10 small veins along leading edge	Forewing length less than 10mm. Overall colour of wings in sub-imago is grey	*Paraleptophlebia* spp. *P. cincta* p. 108 *P. werneri* p. 112
	Forewing length greater than 10mm. Overall colour of wings in sub-imago is mottled brown	*Paraleptophlebia submarginata* p. 110

Habrophlebia fusca (Curtis)
Common name: Ditch Dun

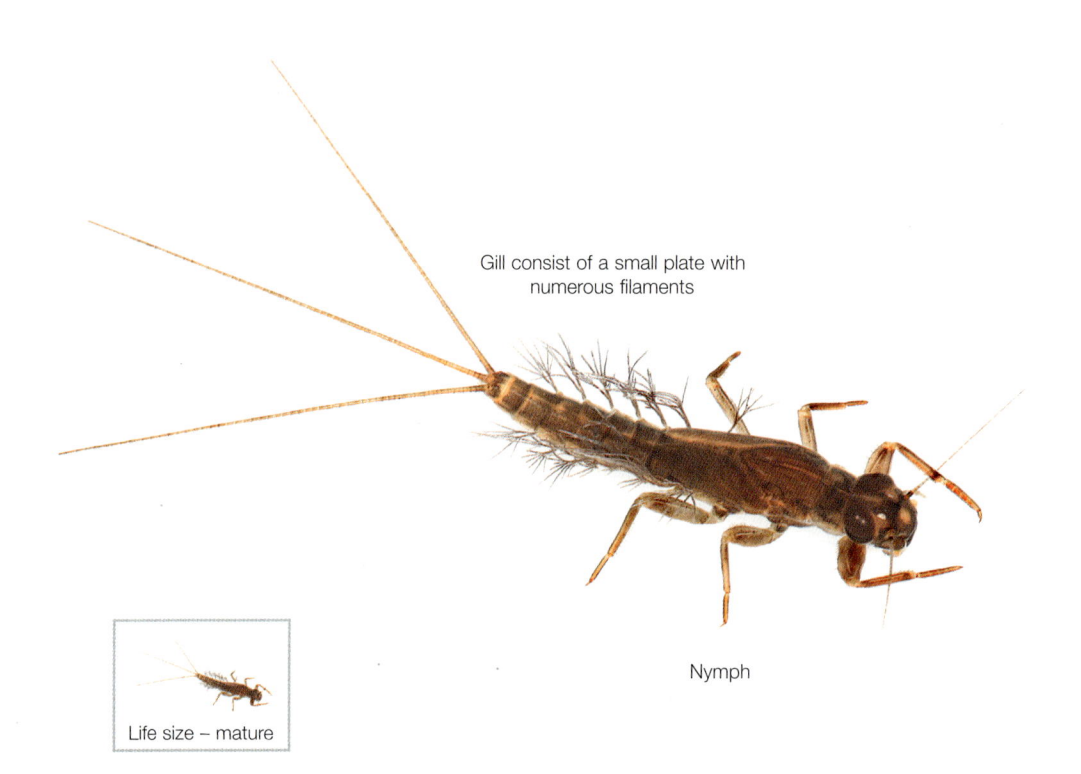

Gill consist of a small plate with numerous filaments

Nymph

Life size – mature

Habrophlebia fusca (Curtis)
Common name: Ditch Dun

Key features

Nymphs: Streamlined nymphs with branched gills. Tails are as long, or longer than the body and held in a 'T' shape at rest.

Adults: Small flies with three tails and large hindwings. The sub-imago has a dark grey body and dark grey wings, whilst the female imago has a reddish-brown body.

Separating from other species

Nymphs: *Habrophlebia fusca* can be separated from the other species of Leptophlebiidae found in the British Isles by the shape of its gills. In *Habrophlebia*, all the gills consist of a small plate with numerous filaments arising from it. Although *H. fusca* is the only species from the genus *Habrophlebia* to be found from the British Isles there are five other species found in mainland Europe. In addition, the European genus *Thraulus* is superficially similar to *Habrophlebia*, but can be separated by the shape of the first gill, which is strap-shaped in *Thraulus* species.

Adults: *H. fusca* can be distinguished from other British Leptophlebiidae by the presence of a small projection on the leading edge of each hindwing. This is similar to the shape of the hindwings in *Serratella ignita*, however in *S. ignita* there are small detached single veins at the edge of the forewings, between each of the major veins. *H. fusca* never has detached veins at the edges of the forewing.

Habitat and Ecology

Nymphs of this species live chiefly in the pools and margins of rivers and stream, where they are found amongst plant life or amongst accumulations of dead leaves. They live on the surface of leaves of aquatic plants or in the surface layers of fine sediments, especially mud where they feed either by filtering or gathering fine particulate organic detritus from the sediment. The nymphs are poor swimmers but are adapted for moving amongst dense stands of plants, especially on the surface of the stems. There is one generation per year, which overwinters as nymphs and emerges between May and September.

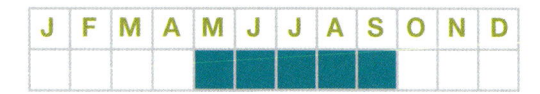

J	F	M	A	M	J	J	A	S	O	N	D
				■	■	■	■	■			

Emergence of the adults typically takes place partially or entirely out of the water on a stick, stones or plant stem during daylight hours. Once mated, the female flies upstream and descends to the surface of the water to release a few eggs by dipping the tip of her abdomen on to the surface at intervals, or by actually settling on the surface for short periods. After several visits to the water the egg supply of around 1,200 eggs is finished and the spent female falls on to the surface.

Distribution

H. fusca is common in central England but has a scattered, localised distribution throughout mainland Britain. There are few records north of the central belt of Scotland and it appears to be absent from the north-west of Scotland.

Leptophlebia marginata (Linnaeus)
Common name: Sepia Dun

Life size

Female sub-imago

The junction of each gill plate with its terminal filament is abrupt

Teeth cover only two thirds of the length of the claw

Spines on lower edge of fore leg have a single point

Life size – mature

Nymph

Leptophlebia marginata (Linnaeus)
Common name: Sepia Dun

Key features

Nymphs: Streamlined nymphs with branched gills. Tails are as long, or longer than the body and held in a 'T' shape at rest.

Adults: Small to medium sized flies with three tails and large hindwings. The sub-imago of *Leptophlebia marginata* has pale fawn, heavily veined wings and an overall dark brown appearance. On closer examination, the body can be seen to be a rich, dark sepia colour. Both the male and the female imago have reddish-brown bodies, however in the male the body segments are ringed with a straw-colour. In both sexes the wings have pale brown veins, however the female imago has smoky black patches along the leading edges of the forewings.

Separating from other species

Nymphs: *Leptophlebia marginata* can be separated from *L. vespertina* by the coverage of teeth on the claws and by the shape of the gills which is more rounded than in *L. vespertina*. *L. marginata* is the larger of the two *Leptophlebia* species found in the British Isles.

Adults: The British *Leptophlebia* species can be separated by the colour and venation of the wings. In *L. marginata* the forewings of the sub-imago are brownish-grey and the veins are marked out in a brownish colour. The hindwings are of a similar shade to the forewings. In *Leptophlebia vespertina* the forewings of the sub-imago are pale grey and the hindwings are noticeably paler than the forewings.

Imagines of *L. marginata* have brownish forewings, particularly at the wing tip and the veins are brown, tending towards yellow. The wings of *L. vespertina* imagines are clear with pale brown veins. There are no appreciable differences in the shade of the hindwings in the imago of either species.

Habitat and Ecology

Nymphs of this species can be found in the pools and margins of slow flowing streams and in ponds and lakes where they climb upon the surface of leaves of aquatic plants or crawl in the surface layers of fine sediments, especially mud. The nymphs are poor swimmers but are adapted for moving amongst dense stands of plants, especially on the surface of the stems. They feed by gathering fine particulate organic detritus from the sediment. There is one generation a year, which usually overwinters as nymphs and emerges between April and June.

J	F	M	A	M	J	J	A	S	O	N	D
			■	■							

Nymphs are seldom found in the shallower margins of lakes in any quantity until early April. As the period of peak emergence approaches, many of the nymphs move into very shallow water. Emergence of the adults takes place during daylight hours at the surface of the water or more typically, partially or entirely out of the water on a stick, stone or plant stem. The males of this species can be found swarming throughout the day.

Once mated, the female flies to the water to release a few eggs by dipping the tip of her abdomen on to the surface at intervals, or by actually settling on the surface for short periods. After several visits to the water her supply of around 1,200 eggs is finished and the spent female falls on to the surface.

Distribution

Leptophlebia marginata is a fairly common species that has been found throughout the British Isles, including Ireland. It is particularly tolerant of the effects of acidification and has been found in waters with pH values between 4 and 5.

Leptophlebia vespertina (Linnaeus)
Common name: Claret Dun

Life size

Female sub-imago

The junction of each gill plate and its terminal is gradual

Teeth cover nearly all the length of the claw

Spines on lower edge of femur of foreleg have many sharp projections

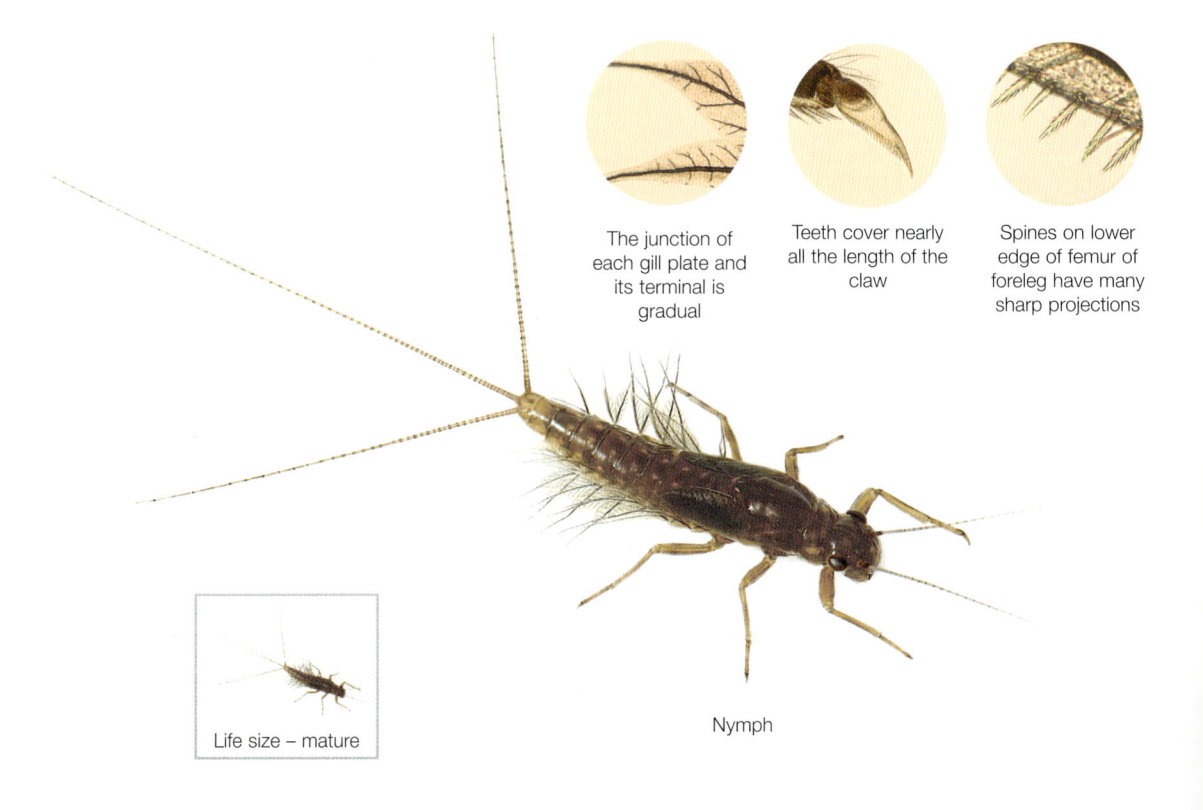

Life size – mature

Nymph

Leptophlebia vespertina (Linnaeus)
Common name: Claret Dun

Key features

Nymphs: Streamlined nymphs with branched gills. Tails are as long, or longer than the body and held in a 'T' shape at rest.

Adults: Small to medium sized flies with three tails and large hindwings. The sub-imago has dark grey forewings and much paler hindwings, particularly in the female. The body appears almost black, but on closer examination can be seen to be dark brown with hints of claret. The eyes of the male are dark red-brown, whilst in the female the eyes are almost black. Both the male and the female imago have brown bodies, which have a distinct tinge of claret to them. The wings in both the male and the female are transparent with pale brown veins.

Separating from other species

Nymphs: *Leptophlebia vespertina* can be separated from *L. marginata* by the coverage of teeth on the claws and by the shape of the gills. The gill plate gradually tapers from the plate to the tip of the gill.

Adults: The British *Leptophlebia* species can be separated by colour and venation of the wings. In *Leptophlebia vespertina* the forewings of the sub-imago are pale grey and the hindwings are noticeably paler than the forewings. In *L. marginata* the forewings of the sub-imago are brownish-grey and the veins are marked out in a brownish colour. The hindwings are of a similar shade to the forewings.

Imagines of *L. vespertina* have clear wings with pale brown veins. The wings of *L. marginata* imagines are brownish, particularly at the wing tip and the veins are brown, tending towards yellow. There are no appreciable differences in the shade of the hindwings in the imago of either species.

Habitat and Ecology

Nymphs of this species can be found in the pools and margins of slow flowing streams and in ponds and lakes where they climb upon the surface of leaves of aquatic plants or crawl in the surface layers of fine sediments, especially mud. The nymphs are poor swimmers but are adapted for moving amongst dense stands of plants, especially on the surface of the stems. They feed by gathering fine particulate organic detritus from the sediment. There is one generation a year, which usually overwinters as nymphs and emerges between April and August.

J	F	M	A	M	J	J	A	S	O	N	D
			■	■	■	■	■				

Emergence of the adults takes place during daylight hours either at the surface of the water, or more typically partially or entirely out of the water on a stick, stone or plant stem. The males of this species can be found swarming throughout the day, and as the species name suggests, this swarming often continues well into the evening.

Once mated, the female returns to the water to release a few eggs by dipping the tip of her abdomen on to the surface at intervals, or by actually settling on the water surface for short periods. After several visits to the water her supply of around between 1,200 and 2,500 eggs is finished and the spent female falls on to the surface.

Distribution

Leptophlebia vespertina is a fairly common species which has been found throughout the British Isles, including Ireland. *L. vespertina* is reported to prefer peaty or acidic waters and as a result, tends to be less common in lowland waters.

Paraleptophlebia cincta (Retzius)
Common name: Purple Dun

Life size

Female sub-imago

Paraleptophlebia cincta (Retzius)
Common name: Purple Dun

Key features

Nymphs: Nymphs with strap-like gills, with the first pair the same size as the others. Teeth on the claw cover just over half the length of the claw.

Adults: Small to medium sized flies with three tails and large hindwings. The male and female sub-imago both have dark brown bodies and dark grey wings, however the male has reddish-brown eyes whilst the females eyes are dull green. The wings of the imago are clear and have pale brown venation. The male imago has a translucent body, the last three segments of which are purple-brown. The female imago has a brownish body that is often tinged with purple. In both the male and the female the tails are white or yellowish.

Separating from other species

Nymphs: *Paraleptophlebia cincta* can be separated from the other British *Paraleptophlebia* species by the coverage of teeth on the claws and by microscopic examination of the spines on the femur. In addition, the first gill is the same size as the rest of the gills whereas in *P. submarginata* the first gill is much smaller than the other gills.

Adults: Adults of *Paraleptophleba cincta* are almost identical to those of *P. werneri*, however they can be separated by the colour of the wings. In *P. cincta* the sub-imago has dark grey wings and the imago has clear wings with indistinct venation. In *P. werneri* the wings of the sub-imago are pale grey and those of the imago are smoky brown with distinct brown veins.

Habitat and Ecology

The nymphs live chiefly in the pools and margins of rivers and streams, where they burrow into, and live in gravel, sand or mud on the bed of the watercourse or forage on moss-covered stones. They are poor swimmers but are adapted for burrowing and moving on moss-covered stones. They feed by filtering or gathering fine particulate organic detritus from the sediment.

There is one generation a year, which usually overwinters as nymphs, although *P. cincta* may also overwinter as eggs. Emerging adults are found between May and August.

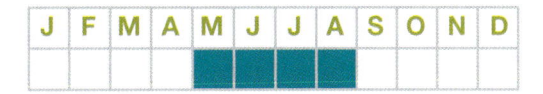

J	F	M	A	M	J	J	A	S	O	N	D
				■	■	■	■				

Once mated, the female flies upstream and descends to the surface of the water where she releases her eggs in a single batch by dipping the tip of her abdomen into the water. After releasing her eggs the spent female falls on to the surface.

Distribution

Paraleptophlebia cincta has been found from near Wick in the north of Scotland to the Lizard peninsula in Cornwall and also from Ireland. Despite its wide distribution there are no large concentrations of records.

Paraleptophlebia submarginata (Stephens)
Common name: Turkey Brown

Life size

Female sub-imago

All gills consist of long narrow filaments. The length of the first gill is much shorter than the second

Life size – mature

Nymph

Paraleptophlebia submarginata (Stephens)
Common name: Turkey Brown

Key features

Nymphs: Nymphs with strap-like gills, with the first pair much smaller than the others. Teeth on the claw cover just over half the length of the claw.

Adults: Small to medium sized flies with three tails and large hindwings. The sub-imago has mottled, fawn-coloured wings with very heavy venation. The forewings have a distinctive clear patch in the centre of the wing. The female has a dark brown body whereas the body of the male is almost black. Both the male and the female imago have brown bodies, however in the male the body is a pale translucent colour with the last three segments of a darker shade. In both sexes the wings have pale brown veins.

Separating from other species

Nymphs: *Paraleptophlebia submarginata* is the only British *Paraleptophlebia* species where the first gill is markedly smaller than the second gill. In the other species the first and second gills are of a similar size.

In *P. submarginata* the teeth on the claw cover just over half the length of the claw. This is also the case in *P. cincta*, however the relative size of the first and second gills allows the separation of these species. In *P. werneri* the teeth on the claw cover about three-quarters of the length of the claw.

In early instars of *Leptophlebia* species (where the body length is less than 3mm), the gills are also strap-like, however the mouthparts can be used to separate the two genera.

Adults: Sub-imagines of *Paraleptophlebia submarginata* are very distinctive. The wings are a mottled fawn colour and have very heavy venation. There are no other British mayflies with three tails, large hindwings and such mottled wings. The only possible confusion could be with *Rhithrogena germanica* which has very similar patterning on the wings, but *R. germanica* only has two tails and is much larger than *P. submarginata*. Unfortunately identification of imagines is more difficult and necessitates microscopic examination of the male genitalia.

Habitat and Ecology

Nymphs of this species live chiefly in the pools and margins of rivers and stream, where they burrow into, and live in gravel, sand or mud on the bed of the watercourse or forage on moss-covered stones. The nymphs are poor swimmers but are adapted for burrowing and moving on moss-covered stones. They feed by filtering or gathering fine particulate organic detritus from the sediment. There is one generation a year, which usually overwinters as nymphs and emerges between April and July.

J	F	M	A	M	J	J	A	S	O	N	D
			■	■	■	■					

Emergence of the adults typically takes place partially or entirely out of the water on a stick, stone or plant stem during daylight hours. The males of this species can be found swarming throughout the day, and often swarming continues until dusk.

Once mated, the female flies upstream and descends to the surface of the water where she releases around 1,200 eggs in a single batch by dipping the tip of her abdomen into the water. After releasing her eggs the spent female falls on to the surface.

Distribution

Paraleptophlebia submarginata is found throughout the British Isles, but has not been recorded from Ireland. The distribution is strongest in central England and Wales but more localised populations are also found as far north as Thurso. Wherever this species is found it is rarely present in large numbers – thorough searching of a site may only reveal a handful of specimens.

Paraleptophlebia werneri Ulmer
Common name: Scarce Purple Dun

Life size

Female sub-imago

Spines on lower edge of femur of hind leg are tapering with pointed tips

Teeth cover three-quarters of the length of the claw

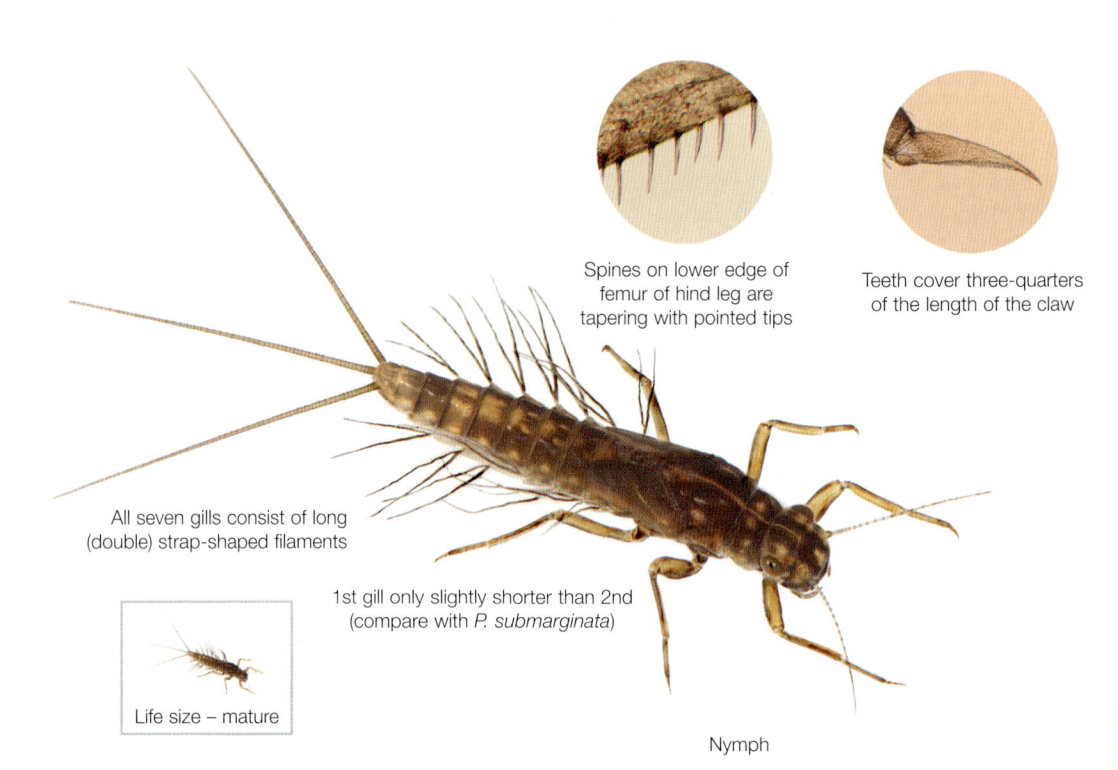

All seven gills consist of long (double) strap-shaped filaments

1st gill only slightly shorter than 2nd (compare with *P. submarginata*)

Life size – mature

Nymph

Paraleptophlebia werneri Ulmer
Common name: Scarce Purple Dun

Key features

Nymphs: Nymphs with strap-like gills, with the first pair the same size as the others. Teeth on the claw cover three-quarters of the length of the claw.

Adults: Small to medium sized flies with three tails and large hindwings. The sub-imago of *Paraleptophlebia werneri* has a dark brown body and light grey wings. The wings of the imago are smoky brown and have pale brown venation. The male imago has a translucent body, the last three segments of which are purple-brown. The female imago has a brownish body that is often tinged with purple. In both the male and the female the tails are greyish-brown.

Separating from other species

Nymphs: *Paraleptophlebia werneri* can be separated from the other British *Paraleptophlebia* species by the coverage of teeth on the claws and by microscopic examination of the spines on the femur. In addition, the first gill is the same size as the rest of the gills whereas in *P. submarginata* the first gill is much smaller than the other gills. Early instars of *Leptophlebia marginata* are often mis-identified as *P. werneri*. To ensure the correct separation of these species detailed microscopic examination of the mouthparts must be undertaken.

Adults: Adults of *Paraleptophlebia werneri* are almost identical to those of *P. cincta*, however they can be separated by the colour of the wings. In *P. werneri* the wings of the sub-imago are pale grey and those of the imago are smoky brown with distinct brown veins. In *P. cincta* the sub-imago has dark grey wings and the imago has clear wings with less distinct venation.

Habitat and Ecology

Nymphs of this species live chiefly in the pools and margins of calcareous watercourses where they burrow into, and live in gravel, sand or mud on the bed of the watercourse or move amongst aquatic plants. This species is often found in watercourses that cease to flow in the summer or are choked with vegetation. The nymphs are poor swimmers but are adapted for burrowing and moving amongst dense stands of plants, especially on the surface of the stems. The nymphs feed by filtering or gathering fine particulate organic detritus from the sediment.

There is one generation a year which usually overwinters as nymphs, but may also overwinter as eggs. Adults can be found during May and June.

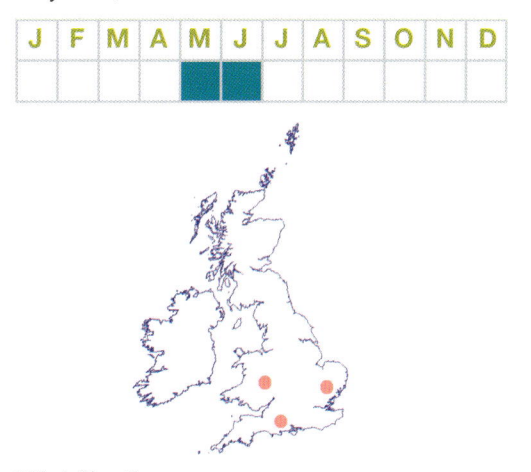

J	F	M	A	M	J	J	A	S	O	N	D
				■	■						

Distribution

Paraleptophlebia werneri is known from very few sites. It was first recorded from Wiltshire in 1939 and since then has been also found in several other river systems in the south of England.

Potamanthus luteus (Linneaus)
Common name: Yellow Mayfly

Life size

Male genitalia (Ventral)

Eyes of newly emerged male

Head of female

Male sub-imago

First gill very small (x 30)

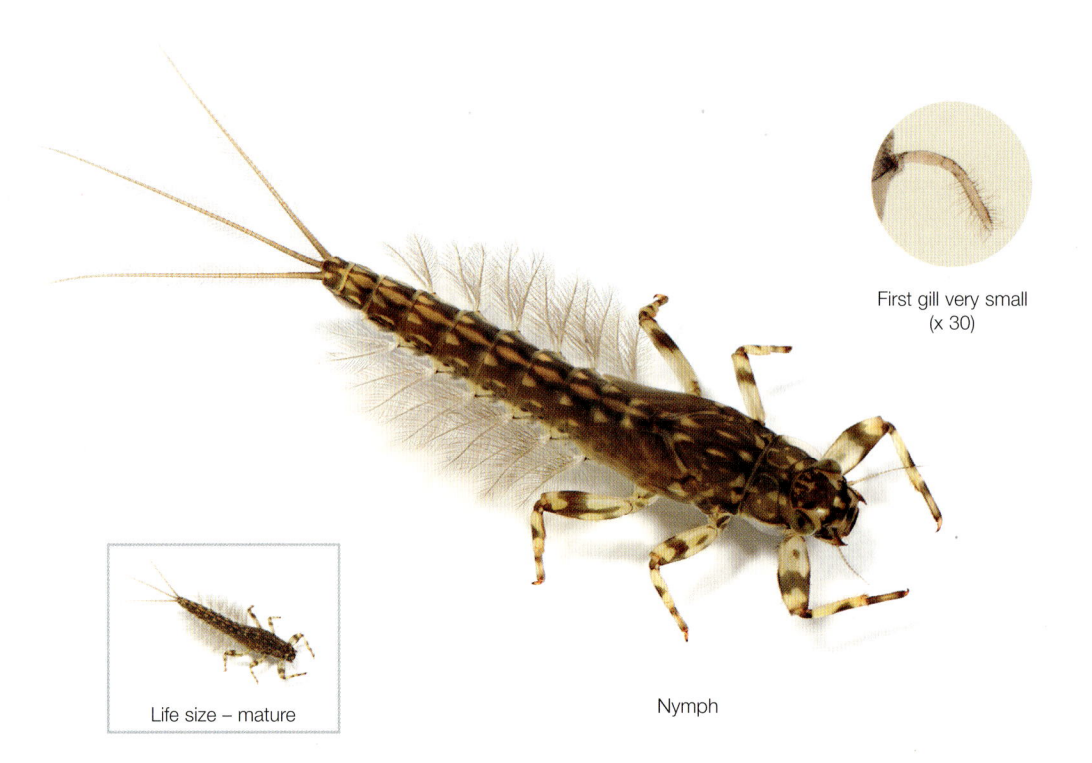

Life size – mature

Nymph

Potamanthus luteus (Linneaus)
Common name: Yellow Mayfly

Key features

Nymphs: Streamlined nymphs with thick, feathery gills that are held at the sides of the body.

Adults: Large flies with three tails and large hindwings. Male sub-imagines of this species have a dull yellowish-orange body with a distinctive broad yellowish-brown stripe along the back of the body. The body is marked with a pair of pale lines and dots on the upper surface of each segment and a single dark dot on the side of each segment. The wings are dull yellow and the cross-veins are a dark reddish colour, particularly at the wing tips. The tails are brown and become progressively lighter further from the body. The eyes in both sexes are yellowish-green.

The imago is similar to the sub-imago, however the wings are brighter yellow and the veins vary between dull yellow to greyish yellow. The tails are yellowish with dark rings and the eyes are yellowish-olive. The female is similar to the male, albeit a slightly brighter shade of yellow and the eyes are dark brown.

Separating from other species

Nymphs: *Potamanthus luteus* has very distinctive yellowish nymphs, with branched feathery gills. The only other species with similar feathery gills are the *Ephemera*, however in *Ephemera* species the gills are held over the back of the body while in *Potamanthus* they extend outwards from the body. In addition, in *Ephemera* species the mandibles are large and project past the front of the head.

Adults: *Potamanthus luteus* is a very distinctive fly. The sub-imago is bright yellow, a characteristic which is shared with only two other species: *Heptagenia sulphurea* and *H. longicauda*.

P. luteus can be separated from the *Heptagenia* species by the number of tails present. In *Heptagenia* there are only two tails while in *P. luteus* there are three.

Habitat and Ecology

Nymphs of this species live chiefly in the pools and margins of larger rivers. They are poor swimmers that typically live amongst stones and sand in side pools. The nymphs feed by gathering fine particulate organic detritus from the sediment.

There is one generation a year, which overwinters as nymphs and emerges principally between May and July, although adults can be found emerging through until early October in some years.

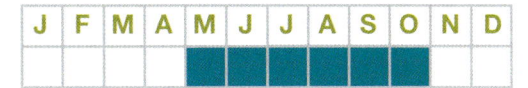

J	F	M	A	M	J	J	A	S	O	N	D
				■	■	■	■	■	■		

Emergence of the adults typically takes place on the surface of the water or partially or entirely out of the water on a stick, stone or plant stem at dusk.

Distribution

Potamanthus luteus is a very rare species with most recent records from the Rivers Usk and Wye in Herefordshire. It is thought that this species has since been lost from the River Usk and has recently suffered a population crash on the River Wye. A voucher specimen is required for all records of this species.

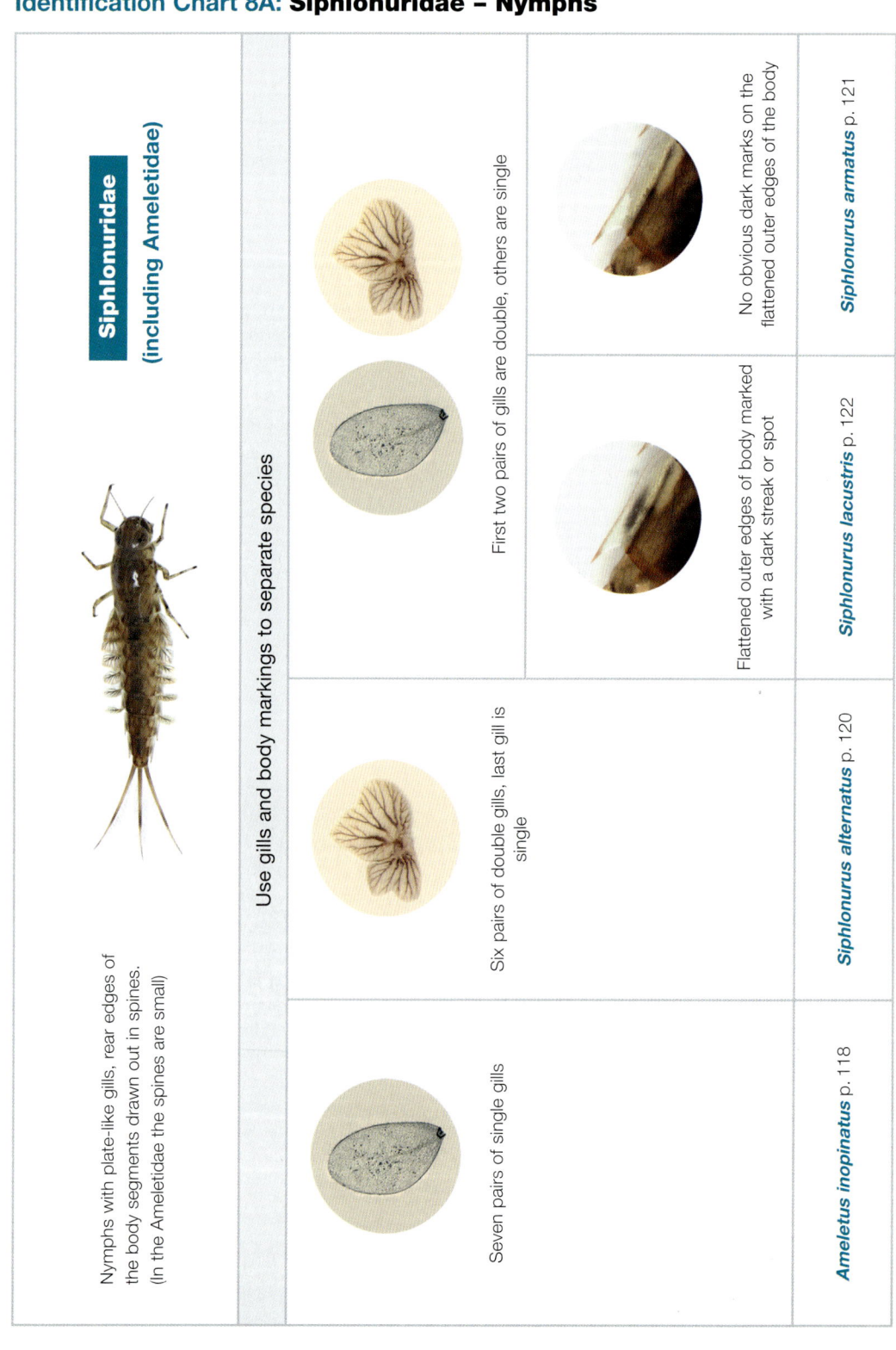

Identification Chart 8A: Siphlonuridae – Nymphs

Siphlonuridae (including Ameletidae)

Nymphs with plate-like gills, rear edges of the body segments drawn out in spines. (In the Ameletidae the spines are small)

Use gills and body markings to separate species

- Seven pairs of single gills
  - *Ameletus inopinatus* p. 118

- Six pairs of double gills, last gill is single
  - *Siphlonurus alternatus* p. 120

- First two pairs of gills are double, others are single
  - Flattened outer edges of body marked with a dark streak or spot
    - *Siphlonurus lacustris* p. 122
  - No obvious dark marks on the flattened outer edges of the body
    - *Siphlonurus armatus* p. 121

Identification Chart 8B: Siphlonuridae – Adults

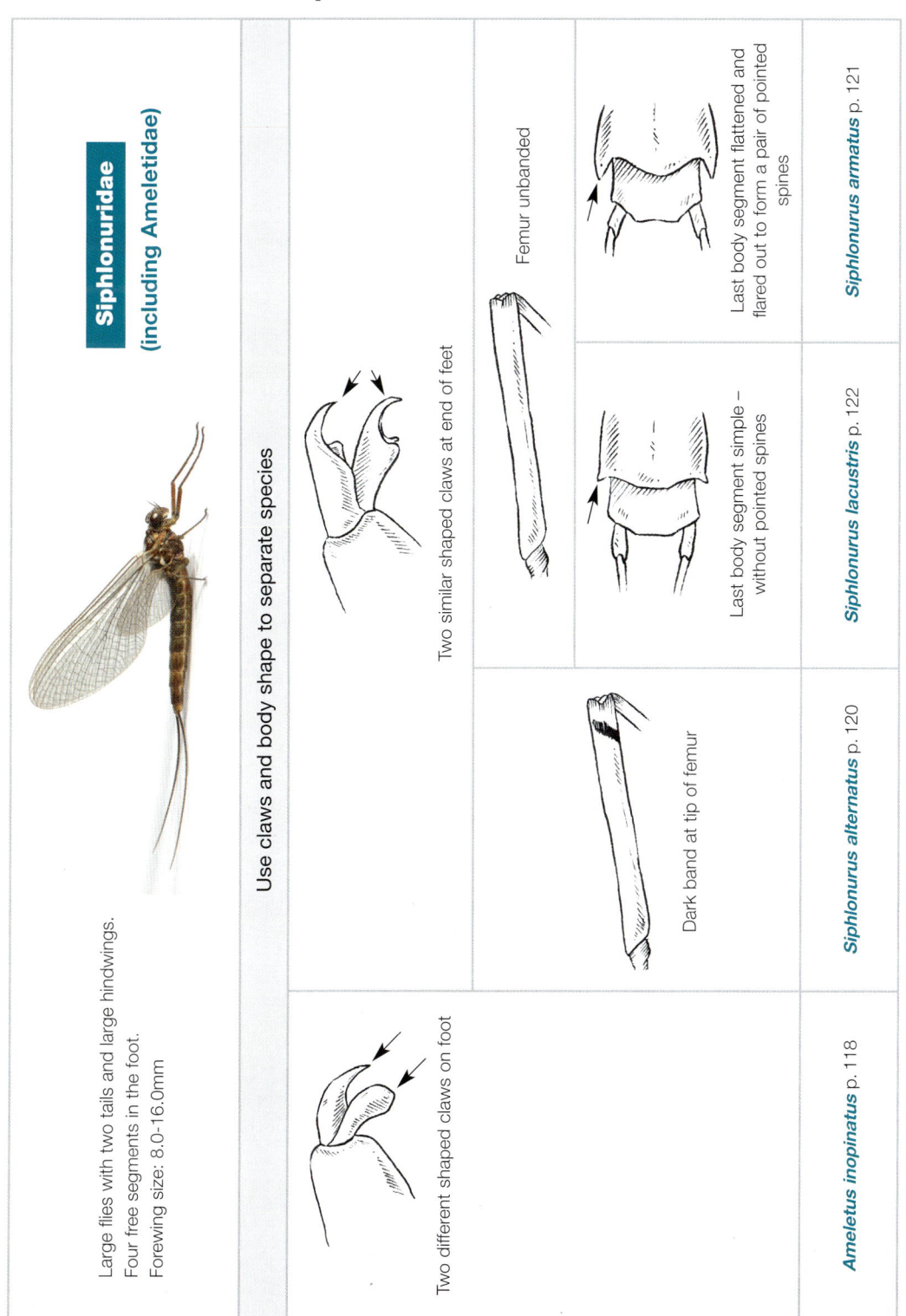

Siphlonuridae (including Ameletidae)

Large flies with two tails and large hindwings.
Four free segments in the foot.
Forewing size: 8.0-16.0mm

Use claws and body shape to separate species

Two similar shaped claws at end of feet

Femur unbanded

Last body segment flattened and flared out to form a pair of pointed spines

Siphlonurus armatus p. 121

Last body segment simple – without pointed spines

Siphlonurus lacustris p. 122

Dark band at tip of femur

Siphlonurus alternatus p. 120

Two different shaped claws on foot

Ameletus inopinatus p. 118

Ameletus inopinatus Eaton
Common name: Upland Summer Mayfly

Life size

Female sub-imago

Spines on hind corners of abdominal segment are small

Maxillae (on mouthparts) have distinctive comb-like bristles

Tails held close together

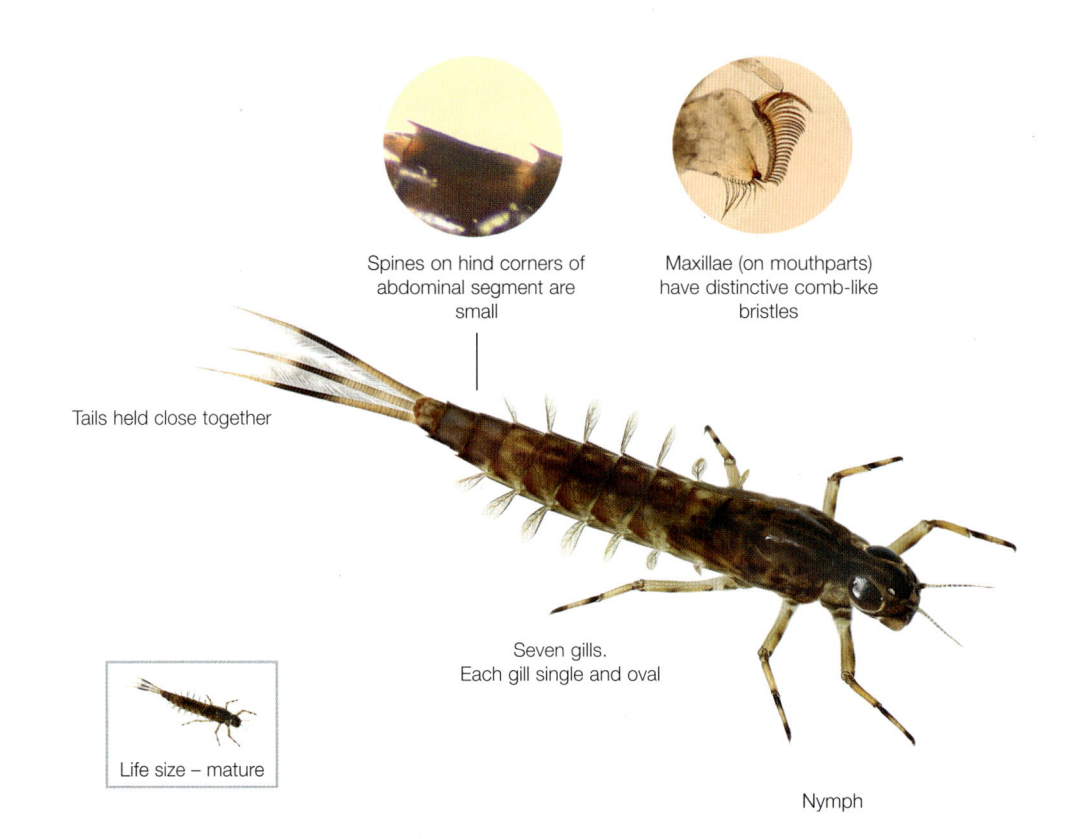

Seven gills. Each gill single and oval

Life size – mature

Nymph

Ameletus inopinatus Eaton
Common name: Upland Summer Mayfly

Key features

Nymphs: Streamlined nymphs with pairs of single plate-like gills. Rear corners of body segments are drawn out into small spines. Tails with a median black band.

Adults: Medium to large flies with two tails and large hindwings. The sub-imago has dark grey wings with yellowish-brown veins. The body is dark brown, with the last few segments often more red in colour. In the imago the wings are predominately clear with a very faint rusty brown sheen and the cross veins are difficult to see. The tails are light brown with almost imperceptible lighter rings to each segment.

Separating from other species

Nymphs: *Ameletus inopinatus* is superficially similar to *Baetis* species, however close examination of the body segments will reveal small spines on the rear corners of each body segment. Caenidae, Heptageniidae and Ephemerellidae also have spines on the body segments however they never have single plate-like gills along their body.

Adults: *Ameletus inopinatus* has two tails and large hindwings in common with the Siphlonuridae and Heptageniidae, however in *A. inopinatus* and *Siphlonurus* species there are four free segments in the foot. In the Heptageniidae all five segments are free. *Ameletus inopinatus* can be separated from *Siphlonurus* species by the claws on each foot. In *A. inopinatus* one of the claws is blunt and round while the other is pointed. In *Siphlonurus* species both claws are pointed.

Habitat and Ecology

Nymphs of this species typically live in upland streams (above 300 metres), but are also found in some Highland lochs in Scotland. The nymphs are good swimmers and typically swim in short, darting bursts. They feed by gathering or collecting fine particulate organic detritus from the sediment.

There is one generation a year, which usually overwinters as nymphs and emerges between May and early August.

J	F	M	A	M	J	J	A	S	O	N	D
				■	■	■	■				

Emergence of the adults typically takes place during daylight hours and males of this species can be found swarming in the afternoon.

Distribution

Ameletus inopinatus has a limited distribution in northern England and Scotland. It is thought that the distribution area of this species will shrink as water temperatures increase due to climate change. A voucher specimen is required for records from all other areas.

Siphlonurus alternatus (Say)
Common name: Northern Summer Mayfly

Key features

Nymphs: Large streamlined nymphs with six pairs of double plate-like gills and a pair of single gills. Rear corners of body segments are drawn out into prominent spines. Tails with a median black band.

Adults: Large flies with two tails and large hindwings. The wings of the sub-imago are grey and the hindwings have a distinctive pale border at the edges. The body, legs and tails are an olive-brown colour with the body having rather variable dark brown markings. In addition, there is a reddish band on each femur. The imago is similar, with an olive-brown body, although in the female this tends to darken to reddish-brown with age.

Separating from other species

Nymphs: *Siphlonurus alternatus* are large nymphs when mature. Immature specimens are superficially similar to *Cloeon* species, however they have large spines projecting from the rear corners of each body segment. Although Heptageniidae, Ephemerellidae and Caenidae also have spines on the body segments, they never have any double plate-like gills along their body.

Siphlonurus alternatus can be separated from other *Siphlonurus* species by the presence of six pairs of double gills and a pair of single gills on the seventh body segment. In *Siphlonurus armatus* and *S. lacustris* only the first two pairs of gills are double, the remainder are single gills.

Adults: Adult Siphlonuridae (the family to which this species belongs) have two tails and large hindwings in common with the Heptageniidae and Ameletidae, however in the Siphlonuridae and Ameletidae there are four free segments in the foot. In the Heptageniidae all five segments are free.

Siphlonurus species can be separated from *Ameletus inopinatus* by the claws on each foot. In *A. inopinatus* one of the claws is blunt and round while the other is pointed. In *Siphlonurus* species both claws are pointed.

Siphlonurus alternatus can be separated from other *Siphlonurus* species by the presence of a dark reddish brown band on the femur just before the joint with the tibia. In addition the ninth body segment is only slightly wider than the tenth segment and in the sub-imago there is a pale outer border to the hindwings.

Habitat and Ecology

Nymphs of this species typically live in deep pools in rivers and streams, but can also be found in calcareous lakes. The large nymphs are good swimmers and typically swim in short, darting bursts. They feed by gathering or collecting fine particulate organic detritus from the sediment. There is one generation a year, which usually overwinters as eggs and emerges between May and August.

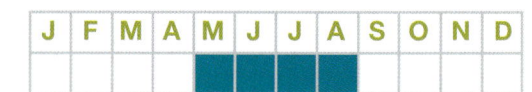

J	F	M	A	M	J	J	A	S	O	N	D
				■	■	■	■				

Emergence of the adults typically takes place during daylight hours and males of this species can be found swarming at dawn and dusk.

Distribution

There are few published records for *Siphlonurus alternatus* from mainland Britain, although it is fairly common in Ireland. As a result, a voucher specimen is required for all records of this species.

Siphlonurus armatus Eaton
Common name: Scarce Summer Mayfly

Key features

Nymphs: Large streamlined nymphs with two pairs of double plate-like gills and five pairs of single gills. Rear corners of body segments are drawn out into prominent spines. Tails with a median black band.

Adults: Large flies with two tails and large hindwings. The wings of the sub-imago are brownish-grey and the cross veins are clearly visible. The body, legs and tails are an olive-brown colour with the body having rather variable dark brown markings. The imago is similar, with an olive-brown body, although in the female this tends to darken to reddish-brown with age.

Separating from other species

Nymphs: *Siphlonurus armatus* are large nymphs when mature. Immature specimens are superficially similar to *Cloeon* species, however they have large spines projecting from the rear corners of each body segment. Heptageniidae, Ephemerellidae and Caenidae also have spines on the body segments, however they never have any double plate-like gills along their body.

Siphlonurus armatus has only two pairs of double gills; the remaining pairs are single gills. It can be separated from other *Siphlonurus* species by the particularly long spines on the ninth body segment. These spines almost reach to the tip of the tenth segment. In addition, the absence of any markings on the flattened edges of the body can be used to separate *S. armatus* however on rare occasions there may be faint dark marks on body segments eight and nine.

Adults: Adult Siphlonuridae (the family to which this species belongs) have two tails and large hindwings in common with the Heptageniidae and Ameletidae, however in the Siphlonuridae and Ameletidae there are four free segments in the foot. In the Heptageniidae all five segments are free.

Siphlonurus species can be separated from *Ameletus inopinatus* by the claws on each foot. In *A. inopinatus* one of the claws is blunt and round while the other is pointed. In *Siphlonurus* species both claws are pointed.

Siphlonurus armatus can be separated from other *Siphlonurus* species by the absence of any markings on the femur and by the shape of the ninth body segment, which is much wider than the tenth segment and markedly flattened at the edges. The ninth segment in *S. lacustris* is slightly wider than the tenth segment but never as noticeably as in *S. armatus*.

Habitat and Ecology

Nymphs of this species typically live in the pools and margins of rivers and streams, or in standing waters. The large nymphs are good swimmers and typically swim in short, darting bursts. They feed by gathering or collecting fine particulate organic detritus from the sediment. There is probably one generation a year, which overwinters as eggs and emerges between May and August.

J	F	M	A	M	J	J	A	S	O	N	D
				■	■	■	■				

Emergence of the adults typically takes place during daylight hours and males of this species can be found swarming at dawn and dusk.

Distribution

There are few published records *Siphlonurus armatus*. As a result, a voucher specimen is required for all records of this species.

Siphlonurus lacustris Eaton
Common name: Summer Mayfly

Life size

Female sub-imago

Large distinct spines and dark marks on hind corners of abdominal segments

First two gills are double plates and overlap which gives a characteristic appearance

Life size – mature

Nymph

Siphlonurus lacustris Eaton
Common name: Summer Mayfly

Key features

Nymphs: Large streamlined nymphs with two pairs of plate-like gills and five pairs of single gills. Rear corners of body segments are drawn out into prominent spines. Tails with a median black band.

Adults: Large flies with two tails and large hindwings. The wings of the sub-imago vary between greyish-brown through to greyish-green. The main veins are dark, however the cross veins are indistinct. The body, legs and tails are an olive-brown colour with the body having rather variable dark brown markings. The imago is similar, with an olive-brown body, although in the female this tends to darken to reddish-brown with age.

Separating from other species

Nymphs: *Siphlonurus lacustris* are large nymphs when mature. Immature specimens are superficially similar to *Cloeon* species, however they have large spines projecting from the rear corners of each body segment. Although Heptageniidae, Ephemerellidae and Caenidae also have spines on the body segments, they never have any double plate-like gills along their body.

Siphlonurus lacustris has only two pairs of double gills; the remaining pairs are single gills. It can be separated from other *Siphlonurus* species by the presence of long dark ovals on the edges of body segments 8 and 9. *S. alternatus* also has dark markings along the edges of the body, however these typically occur on all body segments and are much darker and more square in shape. In addition *S. alternatus* has six pairs of double gills.

Adults: Adult Siphlonuridae (the family to which this species belongs) have two tails and large hindwings in common with the Heptageniidae and Ameletidae, however in the Siphlonuridae and Ameletidae there are four free segments in the foot. In the Heptageniidae all five segments are free.

Siphlonurus species can be separated from *Ameletus inopinatus* by the claws on each foot. In *A. inopinatus* one of the claws is blunt and round while the other is pointed. In *Siphlonurus* species both claws are pointed.

Siphlonurus lacustris can be separated from other *Siphlonurus* species by the absence of any markings on the femur and the width of the ninth body segment, which is only slightly wider than the tenth segment.

Habitat and Ecology

Nymphs of this species typically live in the pools and margins of rivers and streams, lakes, and pools at high altitude. The large nymphs are good swimmers and typically swim in short, darting bursts. They feed by gathering or collecting fine particulate organic detritus from the sediment.

There is one generation a year, which overwinters as eggs and emerges between May and September, although the main flight period is from June to early August.

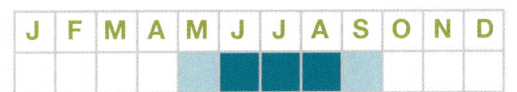

J	F	M	A	M	J	J	A	S	O	N	D

Emergence of the adults probably takes place partially or entirely out of the water on a stick, stone or plant stem during daylight hours. Males of this species can be found swarming throughout the day, including at dawn and dusk. Once mated, the female returns to the water to release a few eggs by dipping the tip of her abdomen into the water at intervals, or by actually settling on the surface for short periods. After several visits to the water her egg supply of upto 2,500 eggs is finished and the spent female falls on to the surface.

Distribution

Siphlonurus lacustris is the most common of the British Siphlonuridae. It occurs in localised pockets throughout the British Isles, including Ireland.

Adult fly identification and flight period tables

3 tails/large hindwings

Species	Common name	Page	J	F	M	A	M	J	J	A	S	O	N	D
Ephemera danica	Green Drake Mayfly	62				●	●	●	●	●	●	●	●	
Ephemera lineata	Striped Mayfly	64							●					
Ephemera vulgata	Drake Mackerel Mayfly	66					●	●	●	●				
Potamanthus luteus	Yellow Mayfly	115					●	●	●	●			●	
Habrophlebia fusca	Ditch Dun	102					●	●	●	●				
Leptophlebia marginata	Sepia Dun	104				●	●	●						
Leptophlebia vespertina	Claret Dun	106				●	●	●						
Paraleptophlebia cincta	Purple Dun	108					●	●	●	●				
Paraleptophlebia submarginata	Turkey Brown	110				●	●	●						
Paraleptophlebia werneri	Scarce Purple Dun	112					●	●	●	●				
Ephemerella notata	Yellow Hawk	71					●	●						
Serratella ignita	Blue Winged Olive	72					●	●	●	●	●			

3 tails/no hindwings

Species	Common name	Page	J	F	M	A	M	J	J	A	S	O	N	D
Brachycercus harrisellus	Large Broadwings	44							●					
Caenis beskidensis	Anglers' Curse	47						●	●	●				
Caenis horaria	Anglers' Curse	48					●	●	●	●				
Caenis luctuosa	Anglers' Curse	50						●	●					
Caenis macrura	Anglers' Curse	52						●	●	●				
Caenis pseudorivulorum	Anglers' Curse	54						●	●	●	●	●		
Caenis pusilla	Anglers' Curse	55						●	●					
Caenis rivulorum	Anglers' Curse	56					●	●	●	●				
Caenis robusta	Anglers' Curse	59						●	●	●				

2 tails/no hindwings

Species	Common name	Page	J	F	M	A	M	J	J	A	S	O	N	D
Cloeon dipterum	Pond Olive	30				●	●	●	●	●	●	●		
Cloeon simile	Lake Olive	34			●	●	●	●	●	●	●	●		
Procloeon bifidum	Pale Evening Dun	38				●	●	●	●	●	●	●		

2 tails/large hindwings

Species	Common name	Page	J	F	M	A	M	J	J	A	S	O	N	D
Siphlonurus armatus	Scarce Summer Mayfly	121					■	■	■	■				
Siphlonurus lacustris	Summer Mayfly	122					■	■	■	■				
Siphlonurus alternatus	Northern Summer Mayfly	120					▨	■	■	■	▨			
Ameletus inopinatus	Upland Summer Mayfly	118					■	■	■	■				
Arthroplea congener	None	99						■	■					
Ecdyonurus dispar	Autumn Dun	76					■	■	■	■	■	■		
Ecdyonurus insignis	Large Green Dun	78					■	■	■	■				
Ecdyonurus torrentis	Large Brook Dun	80				▨	■	■						
Ecdyonurus venosus	Late March Brown	83				■	■	■	■	■	■	■		
Electrogena affinis	Scarce Yellowstreak	95						■	■	■				
Electrogena lateralis	Dusky Yellowstreak	96						■	■	■	■			
Heptagenia longicauda	Scarce Yellow May Dun	86						■	■					
Heptagenia sulphurea	Yellow May Dun	84					■	■	■	■	■	■	■	
Kageronia fuscogrisea	Brown May Dun	89					■	■	■	■				
Rhithrogena germanica	March Brown	90			■	■								
Rhithrogena semicolorata	Olive Upright	92					■	■	■	■				

2 tails/small hindwings

Species	Common name	Page	J	F	M	A	M	J	J	A	S	O	N	D
Baetis atrebatinus	Dark Olive	15					■	■	■	■	■			
Baetis buceratus	Scarce Olive	16						■	■	■				
Baetis digitatus	Scarce Iron Blue	17					■	■	■	■	■			
Baetis fuscatus	Pale Watery	18					■	■	■	■	■			
Baetis muticus	Iron Blue	20				■	■	■	■	■	■	■		
Baetis niger	Southern Iron Blue	22					■	■	■	■	■			
Baetis rhodani	Large Dark Olive	24	■	■	■	■	■	■	■	■	■	■	■	■
Baetis scambus	Small Dark Olive	26			■	■	■	■	■	■	■	■		
Baetis vernus	Medium Olive	28				■	■	■	■	■	■	■		

2 tails/no hindwings

Species	Common name	Page	J	F	M	A	M	J	J	A	S	O	N	D
Centroptilum luteolum	Small Spurwing	36					■	■	■	■	■	■		
Procloeon pennulatum	Large Spurwing	40					■	■	■	■	■			

References

Blackburn, J. H., Gunn, R.J.M. & Hammett, M. J. (1988). *Electrogena affinis* (Eaton, 1885) (Ephemeroptera, Heptageniidae), a mayfly new to Britain. *Entomologist's Monthly Magazine* **134**: 257-263.

Elliott, J.M. & Humpesch, U.M. (1983) A key to the Adults of the British Ephemeroptera with notes on their ecology. *Scientific Publications of the Freshwater Biological Association* No. **47**.

Elliott, J.M. & Humpesch, U.M. (2010). Mayfly Larvae (Ephemeroptera) of Britain and Ireland: Keys and a Review of their Ecology. *Scientific Publications of the Freshwater Biological Association* No. **66**.

Fauna Europaea Checklist (2009). www.faunaeur.org

Gunn, R.J.M. & Blackburn, J.H. (1997). *Caenis pseudorivulorum* Keffermüller (Ephem., Caenidae), a mayfly new to Britain. *Entomologist's Monthly Magazine* **133**: 97-100.

Gunn, R.J.M. & Blackburn, J.H. (1998). *Caenis beskidensis* Sowa (Ephemeroptera, Caenidae), a mayfly new to Britain. *Entomologist's Monthly Magazine* **134**: 94.

Harker, J. (1989). Mayflies. Naturalists' Handbook No. 13.

Teufert, K. (2001). Neuere Untersuchungen und Nachweise von *Heptagenia longicauda* (Stephens, 1835) (Insect: Ephemeroptera) aus dem Leinegebiet der Landkreise Hannover - Hildesheim. *Veröffentlichungen der Gesellschaft für Umweltschutz Langenhagen*.

Useful websites and organisations

Ephemeroptera Recording Scheme
www.ephemeroptera.org.uk

Freshwater Biological Association
The Ferry Landing, Far Sawrey, Ambleside, Cumbria LA22 0LP.
www.fba.org

NBN Gateway
ww.searchnbn.net

Riverfly Partnership
www.riverflies.org

Buglife – The Invertebrate Conservation Trust
www.buglife.org.uk

INDEX TO IDENTIFICATION CHARTS AND SPECIES ACCOUNTS

affinis, Electrogena 95
alternatus, Siphlonurus 120
Ameletidae 116
Ameletus inopinatus 118
armatus, Siphlonurus 121
Arthroplea congener 99
Arthropleidae 74
atrebatinus, Baetis 15

Baetidae 13, 14
Baetis atrebatinus 15
Baetis buceratus 16
Baetis digitatus 17
Baetis fuscatus 18
Baetis muticus 20
Baetis niger 22
Baetis rhodani 24
Baetis scambus 26
Baetis vernus 28
beskidensis, Caenis 47
bifidum, Procloeon 38
Brachycercus harrisellus 44
buceratus, Baetis 16

Caenidae 42, 43
Caenis beskidensis 47
Caenis horaria 48
Caenis luctuosa 50
Caenis macrura 52
Caenis pseudorivulorum 54
Caenis pusilla 55
Caenis rivulorum 56
Caenis robusta 59
Centroptilum luteolum 36
cincta, Paraleptophlebia 108
Cloeon simile 34
Cloeon dipterum 30
congener, Arthroplea 99

danica, Ephemera 62
digitatus, Baetis 17
dipterum, Cloeon 30
dispar, Ecdyonurus 76

Ecdyonurus dispar 76
Ecdyonurus insignis 78
Ecdyonurus torrentis 80
Ecdyonurus venosus 83
Electrogena affinis 95
Electrogena lateralis 96
Ephemera danica 62
Ephemera lineata 64
Ephemera vulgata 66
Ephemerella notata 71
Ephemerellidae 68, 69
Ephemeridae 60, 61

fusca, Habrophlebia 102
fuscatus, Baetis 18
fuscogrisea, Kageronia 89

germanica, Rhithrogena 90

Habrophlebia fusca 102
harrisellus, Brachycercus 44
Heptagenia longicauda 86
Heptagenia sulphurea 84
Heptageniidae 74, 75
horaria, Caenis 48

ignita, Serratella 72
inopinatus, Ameletus 118
insignis, Ecdyonurus 78

Kageronia fuscogrisea 89
lacustris, Siphlonurus 122
lateralis, Electrogena 96
Leptophlebia marginata 104
Leptophlebia vespertina 106
Leptophlebiidae 100, 101
lineata, Ephemera 64
longicauda, Heptagenia 86
luctuosa, Caenis 50
luteolum, Centroptilum 36
luteus, Potamanthus 115

macrura, Caenis 52
marginata, Leptophlebia 104
muticus, Baetis 20

niger, Baetis 22
notata, Ephemerella 71

Paraleptophlebia cincta 108
Paraleptophlebia submarginata 110
Paraleptophlebia werneri 112
pennulatum, Procloeon 40
Potamanthidae 115
Potamanthus luteus 115
Procloeon bifidum 38
Procloeon pennulatum 40
pseudorivulorum, Caenis 54

Rhithrogena germanica 90
Rhithrogena semicolorata 92
rhodani, Baetis 24
rivulorum, Caenis 56
robusta, Caenis 59

scambus, Baetis 26
semicolorata, Rhithrogena 92
Serratella ignita 72
simile, Cloeon 34
Siphlonuridae 116, 117
Siphlonurus alternatus 120
Siphlonurus armatus 121
Siphlonurus lacustris 122
submarginata, Paraleptophlebia 110
sulphurea, Heptagenia 84

torrentis, Ecdyonurus 80

venosus, Ecdyonurus 83
vernus, Baetis 28
vespertina, Leptophlebia 106
vulgata, Ephemera 66

werneri, Paraleptophlebia 112